Lukas Stalder

About nonsense-mediated quality control mechanisms in mammals

Lukas Stalder

About nonsense-mediated quality control mechanisms in mammals

Graduate School for Cellular and Biomedical Sciences

Südwestdeutscher Verlag für Hochschulschriften

Impressum/Imprint (nur für Deutschland/ only for Germany)
Bibliografische Information der Deutschen Nationalbibliothek: Die Deutsche Nationalbibliothek verzeichnet diese Publikation in der Deutschen Nationalbibliografie; detaillierte bibliografische Daten sind im Internet über http://dnb.d-nb.de abrufbar.

Alle in diesem Buch genannten Marken und Produktnamen unterliegen warenzeichen-, marken- oder patentrechtlichem Schutz bzw. sind Warenzeichen oder eingetragene Warenzeichen der jeweiligen Inhaber. Die Wiedergabe von Marken, Produktnamen, Gebrauchsnamen, Handelsnamen, Warenbezeichnungen u.s.w. in diesem Werk berechtigt auch ohne besondere Kennzeichnung nicht zu der Annahme, dass solche Namen im Sinne der Warenzeichen- und Markenschutzgesetzgebung als frei zu betrachten wären und daher von jedermann benutzt werden dürften.

Verlag: Südwestdeutscher Verlag für Hochschulschriften Aktiengesellschaft & Co. KG
Dudweiler Landstr. 99, 66123 Saarbrücken, Deutschland
Telefon +49 681 37 20 271-1, Telefax +49 681 37 20 271-0
Email: info@svh-verlag.de
Zugl.: Bern, Universität Bern, Dissertation 2008

Herstellung in Deutschland:
Schaltungsdienst Lange o.H.G., Berlin
Books on Demand GmbH, Norderstedt
Reha GmbH, Saarbrücken
Amazon Distribution GmbH, Leipzig
ISBN: 978-3-8381-1317-3

Imprint (only for USA, GB)
Bibliographic information published by the Deutsche Nationalbibliothek: The Deutsche Nationalbibliothek lists this publication in the Deutsche Nationalbibliografie; detailed bibliographic data are available in the Internet at http://dnb.d-nb.de.

Any brand names and product names mentioned in this book are subject to trademark, brand or patent protection and are trademarks or registered trademarks of their respective holders. The use of brand names, product names, common names, trade names, product descriptions etc. even without a particular marking in this works is in no way to be construed to mean that such names may be regarded as unrestricted in respect of trademark and brand protection legislation and could thus be used by anyone.

Publisher: Südwestdeutscher Verlag für Hochschulschriften Aktiengesellschaft & Co. KG
Dudweiler Landstr. 99, 66123 Saarbrücken, Germany
Phone +49 681 37 20 271-1, Fax +49 681 37 20 271-0
Email: info@svh-verlag.de

Printed in the U.S.A.
Printed in the U.K. by (see last page)
ISBN: 978-3-8381-1317-3

Copyright © 2010 by the author and Südwestdeutscher Verlag für Hochschulschriften Aktiengesellschaft & Co. KG and licensors
All rights reserved. Saarbrücken 2010

*„In der Wissenschaft gleichen wir alle nur den Kindern,
die am Rande des Wissens hie und da einen Kiesel aufheben,
während sich der weite Ozean des Unbekannten vor unseren Augen erstreckt."*

Isaac Newton

TABLE OF CONTENTS

1. Preamble — 3

2. Introduction and Results — 5
  2.1. Regulatory networks in eukaryotic gene expression — 5
    2.1.1. Complex transcriptional output of a gene locus — 5
  2.2. Surveillance of eukaryotic gene expression — 7
  2.3. Nonsense-mediated mRNA decay (NMD) — 8
    2.3.1. Features and origins of an NMD target — 8
    2.3.2. *Trans*-acting factors involved in NMD — 8
    2.3.3. Subcellular localization of NMD — 9
    2.3.4. PTC recognition — 10
      2.3.4.1. Evidence for the downstream marker model — 10
      2.3.4.2. Evidence for the *faux* 3'UTR model — 11
      2.3.4.3. An evolutionarily conserved model for PTC recognition — 13
    2.3.5. Focusing of UPF1 function — 15
      2.3.5.1. Processing bodies are not required for mammalian NMD — 15
      2.3.5.2. Purification of UPF1 bound PTC+ mRNPs — 18
    2.3.6. Involvement of NMD factors in other cellular processes — 22
    2.3.7. Therapeutic aspects — 22
      2.3.7.1. Suppression of NMD by an anti-sense oligo-nucleotide — 23
      2.3.7.2. A combinatorial approach of PTC124 together with the double-target anti-sense oligo-nucleotide — 26
  2.4. Nonsense mediated transcriptional gene silencing (NMTGS) — 27

3. Discussion and Perspectives — 29
  3.1. The mechanism of NMD is more conserved than previously appreciated — 29
  3.2. Elucidating the function of UPF1 — 30
  3.3. The physiological role of NMD — 31
  3.4. Therapeutic aspects — 32
  3.5. NMTGS – still enigmatic — 34

4. Materials and Methods — 36
  4.1. Plasmids — 36
    4.1.1. Cloning of pSupuro GW182 — 36
    4.1.2. The role of hUpf2 — 36
  4.2. mRNP purifications — 36
    4.2.1. Double-step purification with MS2MBP and α-HA antibodies — 36
    4.2.2. RAT purifications — 37

5. References — 38

6. Appendix — 53
  6.1. Summary — 53
  6.2. Acknowledgments — 54
  6.3. Papers — 55

2

1. Preamble

The publication of Darwins theory "On the Origin of Species by Means of Natural Selection" in 1859 is considered to be the origin of modern biology. Together with the description of the laws of inheritance by the monk Gregor Mendel, with the technical advances of microscopy and with the chemical distinction of organic and inorganic substances, the knowledge about organisms was growing fast at the beginning of the 20th century. A central issue in the history of molecular biology was the description of the double-helical structure of the DNA by Watson and Crick in 1953. 50 years later, in 2003, over 90 % of the human genome was sequenced, heralding a new era in biology. It will be the challenge of the 21st century scientists to decipher the vast amount of genomic sequence.

In my thesis I will first give a short and rather theoretical teaser with the aim to depict the reader the complexity of eukaryotic gene regulatory networks. With the growing complexity of the regulation of eukaryotic gene expression, sophisticated quality-control mechanisms have evolved. During the three years I was working on my thesis in the lab, I mainly focused on the investigation of the molecular mechanisms of two quality-control mechanisms called "Nonsense-mediated mRNA decay" (NMD) and "Nonsense-mediated transcriptional gene silencing" (NMTGS), which are the general focus of the introduction. During the time of my thesis, I was involved in the writing of several review articles, and therefore I allow myself to restrict the introductory paragraphs to facts which either are of interest concerning results of my own or which I find of outstanding interest. Otherwise I will refer to the review articles and I would like to encourage the reader to consult the corresponding paragraphs in the reviews where needed. Furthermore, my results are integrated into the introduction, where they contribute to deeper knowledge. Results that are already described in my publications are only shortly summarized with a reference to the corresponding paper, and results that are of interest but not part of a publication are described in more detail. Finally, in the discussion and perspectives, the results are set into the context of the current knowledge about the quality-control mechanisms NMD and NMTGS.

4

2. INTRODUCTION AND RESULTS

2.1. Regulatory networks in eukaryotic gene expression

The central dogma of molecular biology, the flow of information from DNA to RNA to protein, has been formulated by Francis Crick in 1958. Consequently, it has been assumed that genes mostly code for proteins, which in turn fulfill most structural, enzymatic and regulatory functions in a cell. This is largely true for prokaryotes, whose genomes consist of protein-coding sequences flanked with 5' and 3' cis-acting regulatory regions, although estimates indicate that an eubacterial genome contains small non-coding RNA genes corresponding to 5 % of the number of protein-coding genes (Mattick, 2003; Vogel & Wagner, 2007). The number of regulatory proteins scales almost quadratically with genome size in prokaryotes, suggesting that prokaryotes have been limited in their complexity by the nature of their protein-based regulatory network (van Nimwegen, 2003; Gagen & Mattick, 2005; Mattick & Gagen, 2005). In contrast, beside the expression and regulation of structural and functional components (i.e. proteins and their derived products), the evolution of complex, multi-cellular organisms requires the specification of how these components are spatially and temporally arrayed and assembled into higher levels of organization (i.e. cells and organs), together with systems that control their function. Eukaryotes have breached this limit with the evolution of sophisticated RNA-based gene regulatory networks, in concert with the evolution of the necessary protein infrastructure to recognize and act on these signals (Mattick, 2004). For instance, the evolution of multi-cellular organisms has begun around 1200 million years ago with the entry and expansion of introns into the genome and with the evolution of the spliceosome (Cavalier-Smith, 1991). Concurrently, the number of introns per gene increases with the complexity of the organism (Roy, 2006; Roy & Gilbert, 2006), and introns account for approximately 95 - 97 % of the sequence of human genes. In accordance, although only 2 % of the human genome corresponds to mature protein-coding mRNA sequence (Lander et al., 2001), at least 60 – 70 % of the human genome is transcribed (Carninci et al., 2005; Frith et al., 2005). Many lines of evidence suggest that these non-coding RNAs (ncRNAs), which have been considered as "transcriptional noise", are biologically meaningful and represent the driving force behind the expanding complexity of multi-cellular organisms (Mattick & Makunin, 2006; Mendes Soares & Valcarcel, 2006).

2.1.1. Complex transcriptional output of a gene locus

The transmission of the genetic information encoded on the DNA into a functional protein has turned out to be very complex, involving several multi-component cellular machineries. The pathway of eukaryotic gene expression includes transcription, splicing, capping and polyadenylation, before the mature mRNA can be exported to the cytoplasm, where it can be translated into proteins. Basically every step of processing is tightly regulated, coupled with the other processing steps (Maniatis & Reed, 2002), and is potentially also subject to quality-control (Doma & Parker, 2007) (see also paragraph 2.2.). Furthermore, the use of alternative splicing, alternative promoters and alternative polyadenylation signals significantily contributes to proteomic complexity and provide various opportunities for regulation (Mendes Soares & Valcarcel, 2006)(Fig. 1).

It would go beyond the scope of this introduction to describe transcriptional regulation mechanisms or the regulation of alternative splicing and polyadenylation. However, the existence of those complex processing and regulation events demonstrates the necessity of elaborate quality-control mechanisms in eukaryotic cells, in order to maintain the high accuracy of gene expression. The investigation of two mRNA quality-control mechanisms termed nonsense-mediated mRNA decay (NMD) and nonsense-mediated transcriptional gene silencing (NMTGS) is the topic of my thesis, and therefore I will give a broad introduction into those mechanisms in the subsequent paragraphs.

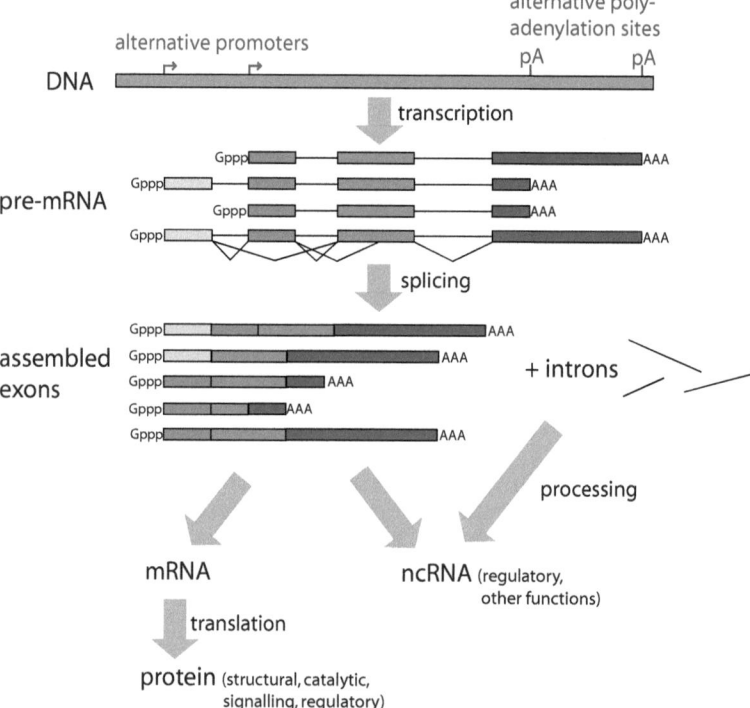

FIGURE 1: **Transcript diversity and combinatorial regulation of eukaryotic gene expression.** Various transcripts can be generated from one gene locus by the use of alternative promoters, alternative polyadenylation sites and alternative splicing. Only few examples of pre-mRNA and assembled exons are shown, theoretical 12 different transcripts could be generated in this case. The stability, translation and the cellular localization of the different isoforms can be regulated by RNA-binding factors. For instance, the region between the poly(A) sites might contain binding sites for miRNAs to regulate the translation of these isoforms. Furthermore, the introns and also the assembled exons can be processed into various types of ncRNA.

2.2. Surveillance of eukaryotic gene expression

Many steps of RNA biogenesis and function, if not all, are subject to quality control (see Fig. 2; reviewed in (Doma & Parker, 2007)). Aberrant RNAs can arise from a variety of sources, and they are usually recognized and targeted for degradation by the RNA quality-control mechanisms. For instance, tRNAs with a defective modification or nonfunctional rRNAs are recognized by yet unknown factors and degraded by cytoplasmatic mechanisms called "rapid tRNA decay" (RTD) or "nonfunctional rRNA decay" (NRD), respectively (Alexandrov et al., 2006; LaRiviere et al., 2006). Furthermore, beside the surveillance mechanisms acting on nuclear RNAs, several mechanisms exist to control the fate of aberrant cytoplasmatic mRNA. Generally, translation is an intrinsic requirement of the cytoplasmic quality-control mechanisms. Nonstop decay (NSD) degrades transcripts that lack a termination codon. In yeast, ribosomes that encounter the end of an mRNA lacking translation termination codons can recruit the exosome via the interaction of Ski7p with the empty ribosomal A site (Frischmeyer et al., 2002; van Hoof et al., 2002). If a ribosome pauses the elongation, for instance due to a stable stem-loop structure, this mRNA is degraded by a mechanism called no-go decay (NGD) (Doma & Parker, 2006). NGD has mainly been studied in *S. cerevisiae*, and it remains to be elucidated if it also occurs in mammals. Furthermore, when ribosomes translate into the 3'UTR and terminate within the 3'UTR, some mRNAs are degraded in a process called ribosome extension-mediated decay (REMD) (Inada & Aiba, 2005; Kong & Liebhaber, 2007). However, the best studied cytoplasmatic mRNA quality-control mechanism is nonsense-mediated mRNA decay (NMD), which recognizes and degrades mRNAs containing a premature translation-termination codon (PTC). Yet another surveillance mechanism called non-

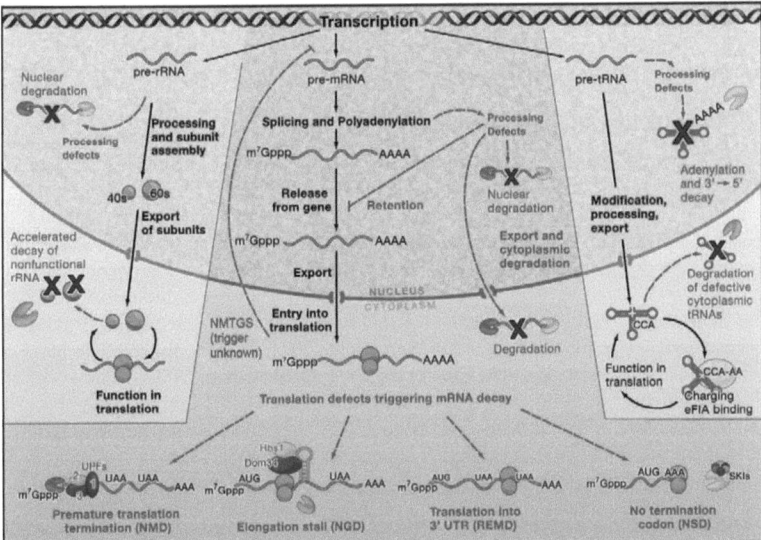

FIGURE 2: **Diversity of RNA quality-control systems in eukaryotes.** The figure depicts some of the known RNA quality-control systems for aberrant rRNA, mRNA, and tRNA in eukaryotic cells (adapted from (Doma & Parker, 2007)).

sense-mediated transcriptional gene silencing (NMTGS) was recently discovered in our lab (Buhler et al., 2005).

2.3. Nonsense-mediated mRNA decay (NMD)

2.3.1. Features and origins of an NMD target

PTCs can arise from mutations at the DNA and at the RNA levels (for details see: paper I and paper V). Mutations of the DNA can either directly create an UAA, UGA or UAG stop codon by a single nucleotide substitution, by random nucleotide insertions or deletions that result in a frame-shift, or by mutations changing splicing signals that lead to aberrant splicing. The transcripts of the immunoglobulin superfamily genes represent a special class of NMD targets that usually undergo very efficient NMD. PTCs occur frequently from the immunoglobulin genes, because two of three rearrangements of the V, D and J segments result in a frameshift (Jack et al., 1989; Connor et al., 1994; Carter et al., 1996; Buzina & Shulman, 1999). In a very low frequency, PTCs can arise from errors during the transcription of the pre-mRNA. In contrast, 45 % of alternatively spliced mRNAs have been predicted to be an NMD target (Lewis et al., 2003). Furthermore, the population of NMD substrates is not only restricted to faulty transcripts, but also comprises numerous endogenous, physiological transcripts. Among those are i) mRNAs containing short upstream ORFs (uORFs) that play a central role regulating translation and turnover, ii) mRNAs encoding selenocysteine-containing proteins, iii) mRNAs harboring introns in their 3'UTR, iv) mRNAs with long 3' UTRs, and v) transposons and retroviruses (Muhlrad & Parker, 1999; Ruiz-Echevarria & Peltz, 2000; Sun et al., 2001; He et al., 2003; Mendell et al., 2004; Rehwinkel et al., 2005). In addition, pseudogenes, bicistronic mRNAs, and mRNAs containing signals for programmed frameshifting have been identified as NMD targets in yeast (He et al., 2003).

2.3.2. *Trans*-acting factors involved in NMD

The *trans*-acting factors involved in NMD are described in-depth in paper V, therefore I will only summarize some of the important features here. The three UPF (Up-frameshift) proteins -UPF1, UPF2 and UPF3- are conserved from yeast to human and were identified as principal NMD-factors in all eukaryotes studied so far. In metazoans, SMG1, SMG5, SMG6 and SMG7 were identified as additional NMD factors (reviewed in (Behm-Ansmant et al., 2007; Chang et al., 2007) and paper V). UPF1 interacts with the eukaryotic release factors eRF1 and eRF3, it binds to UPF2, and it interacts with SMG1, SMG5, SMG6 and SMG7 (Czaplinski et al., 1998; Serin et al., 2001; Wang et al., 2001; Yamashita et al., 2001; Anders et al., 2003; Chiu et al., 2003; Ohnishi et al., 2003; Fukuhara et al., 2005; Kashima et al., 2006; Ivanov et al., 2008). Moreover, UPF1 exhibits RNA binding, RNA-dependent ATP hydrolysis, and 5'-to-3' ATP-dependent RNA helicase activities, and inhibition of any of these activities suppresses NMD (Bhattacharya et al., 2000; Kashima et al., 2006; Cheng et al., 2007). Furthermore, UPF1 undergoes a cycle of phosphorylation and dephosphorylation, which is essential for NMD in metazoans. Phosphorylation of UPF2 is catalyzed by SMG1 and requires UPF2 and UPF3 (Yamashita et al., 2001; Kashima et al., 2006), whereas dephosphorylation of UPF1 is mediated by SMG5, SMG6 and SMG7, which are thought to recruit protein phosphatase 2A (PP2A) (Page

Introduction and Results

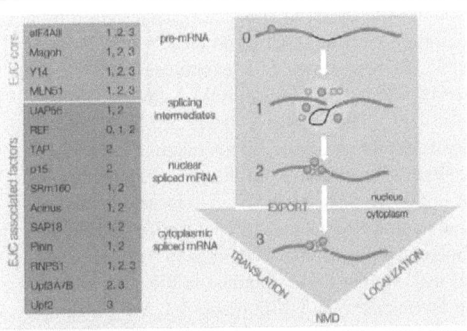

FIGURE 3: **Protein composition of the EJC along the travel of mRNAs from the nucleus to the cytoplasm.** The EJC interferes with several cellular events including mRNA export, translation, localization and nonsense-mediated mRNA decay (NMD). The EJC is depicted in groups of proteins, the EJC core components (yellow) and the EJC associated factors (green). The EJC core components (yellow) form a stable complex on nuclear spliced mRNAs (2) and remain stably associated with these mRNAs after their transport into the cytoplasm (3). More peripheral EJC associated factors (green) are associated to mRNA at different stages of their metabolism: precursor mRNAs (pre-mRNAs, 0), splicing intermediates (1), nuclear (2) or cytoplasmic (3) spliced mRNAs (adapted from (Le Hir & Andersen, 2008)).

et al., 1999; Anders et al., 2003; Chiu et al., 2003; Ohnishi et al., 2003; Kashima et al., 2006).

Furthermore, data from yeast, flies and humans revealed that poly(A) binding protein (PABP) is an NMD antagonizing factor, since NMD can be suppressed by tethering PABP near the PTC (Amrani et al., 2004; Behm-Ansmant et al., 2007; Silva et al., 2008; Singh et al., 2008) (and paper III). Moreover, a protein complex called exon junction complex (EJC), which is deposited about 20-24 nucleotides upstream of exon-exon junctions in spliced mRNA, plays an important role in mammalian NMD (see paragraph 2.3.5.). Y14, MAGOH, eIF4A3, and Barentsz/MLN51 form the core of the EJC and are bound to the mRNA in the nucleus and in the cytoplasm (reviewed in (Tange et al., 2004; Le Hir & Andersen, 2008)). Additional factors associate with EJCs more dynamically and some already leave before mRNA export or only bind the EJC in the cytoplasm. Importantly, UPF2 and UPF3 have been identified as EJC components in mammalian cells (Fig. 3).

2.3.3. Subcellular localization of NMD

The subcellular localization of NMD is still up for debate. Recently, it has been demonstrated that the expression of polypetides, which were designed to inhibit several interactions between NMD factors, only inhibited NMD when expressed exogenously in the cytoplasm, but not when they were confined to the nucleus by means of a nuclear localization signal (NLS) (Singh et al., 2007). These results suggest that most, if not all, NMD occurs in the cytoplasm of mammalian cells. However, the reported evidence for nuclear translation in human cells (Iborra et al., 2001), the occurrence of NMD on CBC-bound mRNA (Ishigaki et al., 2001), and the finding that NMD was not suppressed by inhibition of mRNA export (Buhler et al., 2002) would be consistent with a model for intranuclear NMD. Although this interpretation is mechanistically interesting, current models for NMD do not account for the possibility of intranuclear NMD, and presumably, given that the existence of intranuclear NMD can not be demonstrated directly, it will not be possible to exclude potential intranuclear NMD experimentally.

However, a recent study provides evidence that NMD occurs at least partially in cytoplasmatic processing bodies (P-bodies, also known as "GW182-bodies", "DCP1-foci" or "XRN1-foci") in *S. cerevisiae* (Sheth & Parker, 2006). P-bodies have been identified in both yeast and mammalian cells to be sites of mRNA turnover and storage (Sheth &

Parker, 2003; Cougot et al., 2004). P-bodies are dynamic structures characterized by a high local concentration of mRNA decapping enzyme (DCP1 and DCP2), activators of decapping (Ge-1, EDC3, Lsm1-7, RAP55, RCK/p54), the 5'-3' exonuclease XRN1, the deadenylation-complex CCR4-CAF1-NOT, and factors of the miRNA pathway (GW182, Argonaute proteins) (reviewed in (Eulalio et al., 2007; Parker & Sheth, 2007)). In contrast to the situation in yeast, microscopically detectable P-bodies are not required for NMD in *D. melanogaster* (Eulalio et al., 2007). Otherwise, nonsense mRNAs in mammalian cells were also proposed to be degraded in P-bodies, as SMG7 was found to localize to distinct cytoplasmic foci called processing bodies, and SMG5 and UPF1 localize to P-bodies when SMG7 is overexpressed (Unterholzner & Izaurralde, 2004; Fukuhara et al., 2005; Durand et al., 2007). However, I found that, similar to the situation in flies, microscopically visible P-bodies are not required for mammalian NMD (paper VI, and paragraph 2.3.6.1.).

2.3.4. PTC recognition

A central step in the mechanism of NMD is the distinction of a PTC from a normal translation-termination codon (TC). The proposed models for PTC recognition in the different species studied so far are remarkably different, but they essentially fall into two broad categories (Shyu et al., 2008). The "downstream marker model" proposes that "marker proteins", deposited downstream of a PTC, distinguish a PTC from a normal stop codon. This model has led to the suggestion that a termination codon located upstream of the marker proteins will be recognized as premature, which ultimately leads to the degradation of the mRNA. This model also predicts that the marker proteins bound to the mRNA are displaced during a first round of translation, therefore rendering this mRNA immune to NMD for the subsequent rounds of translation (Fig. 4A). The alternative model for PTC recognition invokes a *"faux* 3'UTR". This model postulates that proper (or efficient) translation termination requires a termination-promoting signal, and that the absence of this signal typifies aberrant translation-termination at a PTC, which in turn leads to degradation of the mRNA (Amrani et al., 2004; Amrani et al., 2006) (Fig. 4B). In the two following paragraphs, I will summarize the experimental evidence for both models, and then I will describe the results from our work, which allow a conjunction of the downstream marker model and the *faux* 3'UTR model into one, unified model for PTC recognition.

2.3.4.1. Evidence for the downstream marker model

Studies on the yeast PTC-containing PGK1 mRNA led to the characterization of a so called downstream sequence element (DSE) in *S. cerevisiae* (Peltz et al., 1993). The PTC-containing PGK1 mRNA was found to be destabilized by a short sequence element (DSE), and vice versa the PGK1 mRNA was stabilized the removal of the DSE (Peltz et al., 1993). Hrp1p was identified as binding partner for the DSE, and Hrp1p is essential for the DSE-mediated NMD of PGK1 (Gonzalez et al., 2000). Thus, Hrp1p can be considered as a downstream marker for NMD on PGK1 mRNA. Although other DSE sequences have been identified, this model may not be generally valid, because of the weak sequence conservation of the DSEs and no DSEs were found in other organisms. Interestingly, also mRNAs containing stabilizer elements (STE) were found, with the ability to render those mRNAs immune to NMD (Ruiz-Echevarria et al., 1998; Ruiz-Echevarria &

Peltz, 2000).

As predicted by the downstream marker model, several lines of evidence suggested that NMD on mammalian transcripts is confined to cap-binding complex (CBC)-bound mRNAs, which led to a model where NMD is restricted to a so called "pioneer round of translation" (reviewed in (Chang et al., 2007; Isken & Maquat, 2007)). Among the few reporter genes studied in mammals, mostly PTCs located 55 nucleotides upstream of the last intron triggered robust NMD (reviewed in (Maquat, 2004; Isken & Maquat, 2007). The discovery that splicing deposits an EJC 20-24 nt upstream of the exon-exon junction led to a model where a PTC was recognized when positioned upstream of the last EJC (Fig. 4A). Furthermore, RNAi-mediated depletion of EJC components reduced the efficiency of NMD (Mendell et al., 2002; Palacios et al., 2004; Gehring et al., 2005; Kim et al., 2005; Buhler et al., 2006; Chan et al., 2007), and tethering of EJC components downstream of a PTC triggered degradation (Lykke-Andersen et al., 2001; Gehring et al., 2003; Palacios et al., 2004). Thus, a 3'UTR bound EJC downstream of a PTC can serve as a downstream marker in mammalian cells. However, although EJC components are conserved in *D. melanogaster* and *C. elegans*, NMD is splicing-independent in these organisms (Gatfield et al., 2003; Behm-Ansmant et al., 2007; Longman et al., 2007), suggesting that NMD is EJC-independent in worms and flies. Additionally, several transcripts studied in mammalian cells can be degraded by NMD without an EJC downstream of the PTC (reviewed in (Chang et al., 2007)), including our Ig-µ minigene, suggesting that the downstream marker model can only partially explain the experimental data.

2.3.4.2. Evidence for the *faux* 3'UTR model

The *faux* 3'UTR model is mainly based on a study performed in *S. cerevisiae* (Amrani et al., 2004), but recent studies performed in worms, flies, plants and mammals provide more evidence for this model (Schwartz et al., 2006; Behm-Ansmant et al., 2007; Longman et al., 2007; Eberle et al., 2008; Ivanov et al., 2008; Singh et al., 2008). Basically, the *faux* 3'UTR model postulates a mechanistic difference between the translation-termination at a normal stop codon and at a PTC (Amrani et al., 2004; Amrani et al., 2006). According to this model, proper translation termination requires the presence of termination promoting signals, which renders an mRNA stable and leads to efficient translation. Otherwise, aberrant translation termination at a PTC would lead to the degradation of the mRNA (Fig. 4B). The exact molecular events of translation termination are not yet solved, but it seems to be clear that for proper termination a certain mRNP structure containing a specific set of factors is required, probably including the possibility for the recycling of the terminating ribosome to the 5' end of the mRNA (Amrani et al., 2004; Amrani et al., 2006). The closed-loop structure of an mRNP (Wells et al., 1998), which is formed after the export through the interaction of Pap1p, eIF4G and eIF4E might represent such a structural environment for proper translation-termination. Consistently, tethering of Pap1p downstream of a PTC rescues the stability of the mRNA in *S. cerevisiae* (Amrani et al., 2004). Furthermore, tethering of PABPC1 (the major cytoplasmic PABP) downstream of a PTC in *drosophila S2* cells and human cells also leads to a stabilization of the mRNA (Behm-Ansmant et al., 2007; Ivanov et al., 2008; Silva et al., 2008; Singh et al., 2008; and paper III). This suggests that PABP can serve as a signal to promote correct translation termination. Otherwise, the *faux* 3'UTR model predicts that the lack of such a translation termination stimulating environment leads to aberrant translation termination, which may

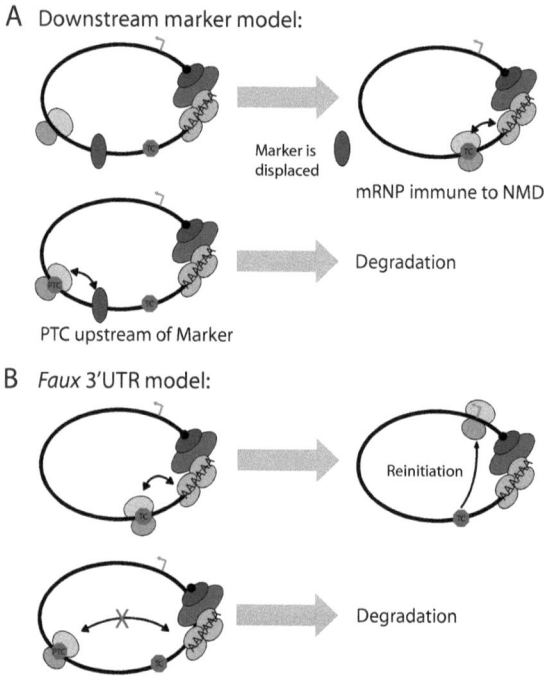

FIGURE 4: **The "Downstream marker model" and the "*faux* 3'UTR model" for PTC recognition.** (A) The mRNP is thought to form a closed-loop structure through the interaction of cap-bound eIF4E or CBC with eIF4G, which in turn interacts with the poly(A) tail-bound PABPC1. The downstream marker model proposes that on a normal mRNA, the bound marker proteins are displaced from the mRNA by the first translating ribosome, rendering this mRNA immune to NMD in subsequent rounds of translation. If the ribosome stops at a PTC, the interaction with the downstream bound marker proteins leads to the degradation of the mRNA. (B) The *faux* 3'UTR model proposes that the ribosome terminates aberrant at a PTC due to the lack of termination promoting signals (i.e. interaction with PABPC1), which leads to the degradation of the mRNA, whereas the mRNA remains stable when the ribosome terminates correct at the TC in close vicinity to the poly(A) tail.

be biochemically different from termination at a normal stop codon. Supporting this idea, toe-prints from ribosomes terminating at a PTC, but not from normal stop codons, could be detected in *S. cerevisiae* (Amrani et al., 2004), indicating that the terminating ribosome stalls at the PTC. Furthermore, a drug called PTC-124 was recently described, which can specifically promote read-through at PTCs, further corroborating the idea that aberrant translation-termination at a PTC is mechanistically different from translation-termination at a normal stop codon. As a consequence of the ribosome stalling at the PTC, the kinetics of aberrant translation-termination might be slow, and the recruitment of Upf1p to the terminating ribosome can be promoted via the interaction of Upf1p with Sup35p and Sup45p, the yeast homologues of eRF3 and eRF1, respectively, which finally leads to the degradation of the mRNA (Czaplinski et al., 1998; Hilleren & Parker, 1999). However, there are also problems with the *faux* 3'UTR model. For instance, the *faux* 3'UTR model does not easily account for the roles of DSEs and EJCs in NMD. Moreover, especially mRNAs in higher eukaryotes often have long 3'UTRs, which makes it difficult to believe of a general important role of 3'UTR-length in NMD.

2.3.4.3. An evolutionarily conserved model for PTC recognition

Recently, our group could show that the EJC is not essential for PTC recognition on our Ig-μ minigene (Buhler et al., 2006). PTCs as close as 10 nucleotides upstream of the last exon-exon junction still elicited NMD, and even PTCs without a downstream EJC were targeted by NMD (Buhler et al., 2004; Buhler et al., 2006). RNAi-mediated depletion of the EJC core factor eIF4AIII led only to a reduced NMD efficiency, but did not abolish NMD (Palacios et al., 2004). This suggests that the EJC is not directly involved in PTC recognition, but may act as an enhancer of NMD (see paragraph 3.1., and papers III and IV). These results are contradictory to the downstream marker model. Strikingly, a Ig-μ reporter with an extension in the 3'UTR had a reduced steady-state mRNA level in a UPF1-dependent manner (Buhler et al., 2006). This suggests that the *faux* 3'UTR model at least partially might also be true in mammals.

Thus, both the downstream marker model and the *faux* 3'UTR model seemed to explain certain aspects of mammalian NMD, and it was therefore important to further investigate in order to find a "unified model", which ideally can explain all observations of mammalian NMD. First, it was important to further evaluate the effect of 3'UTR extensions on NMD reporter genes. The Ig-μ reporter used in Buhler et al., 2006 has an insertion of 24 MS2 binding sites in the 3'UTR. A problem with this reporter is that MS2 binding sites adopt a certain folding, but it is not known, how they are folded when many repeats are present. We therefore designed Ig-μ minigenes containing either one fragment (600 bp long) or two fragments (1200 bp long) of the ampicillin resistance gene in the antisense orientation in the 3'UTR. This prokaryotic sequence was supposed to be unfolded and not to contain an open reading frame or binding sites for eukaryotic RNA binding proteins. Indeed, the steady-state mRNA levels were gradually decreased with increasing length of the 3'UTR, the half-live of the Ig-μ +1200 bp mRNA was reduced compared to the corresponding WT mRNA, and the Ig-μ +1200 bp mRNA could be stabilized by RNAi-mediated depletion of the NMD factors UPF1, UPF2 and UPF3b (paper III). These experiments show that an mRNA with a normal stop codon can be turned into a *bona fide* NMD substrate by extending the 3'UTR in mammalian cells. Furthermore, mRNAs with a long 3'UTR were recently identified as being NMD substrates in *S. cerevisiae, D. melanogaster, C. elegans* and *A. thaliana* (Amrani et al., 2004; Kertesz et al., 2006; Schwartz et al., 2006; Behm-Ansmant et al., 2007; Longman et al., 2007).

Especially higher eukaryotes often have mRNAs with long 3'UTRs, and therefore the simple length in nucleotides from the stop codon to the poly(A) tail could not be the only determinant for PTC recognition. It is likely that many 3'UTRs are structured, and therefore we hypothesized that the physical distance between the stop codon and the poly(A) tail could be a critical determinant for PTC recognition. Consistently, the PTC-containing Ig-μter440 reporter mRNA could be stabilized by folding back the poly(A) tail near downstream of the PTC (paper III). Furthermore, the stabilizing effect gradually decreased with increasing distance of the PTC to the poly(A) tail. Moreover, the Ig-μter310 reporter mRNA, which contains a PTC followed by 2 downstream EJCs, could also be stabilized using the same fold-back approach. These results suggest that the physical distance between the stop codon and the poly(A) tail is the critical determinant for PTC recognition. Additionally, we could show that PABPC1 can suppress NMD when tethered downstream of a PTC, confirming previous results from *S. cerevisiae* and *D. melanogaster* (Amrani et al., 2004; Behm-Ansmant et al., 2007). Moreover, this is consistent with the observation that UPF1 and PABPC1 compete for the interaction with eRF3 in human cells in vitro (Singh et al., 2008).

Based on the data described above and the recent reports from different organism cited above, we propose a unified, evolutionarily conserved model for PTC recognition among all eukaryotes (paper III and paragraph 3.1.). This model essentially extends the *faux* 3'UTR model for PTC recognition to all eukaryotes and introduces EJC and DSEs as enhancers of NMD (Fig. 5). For a detailed description of the unified model for NMD, please see paper IV.

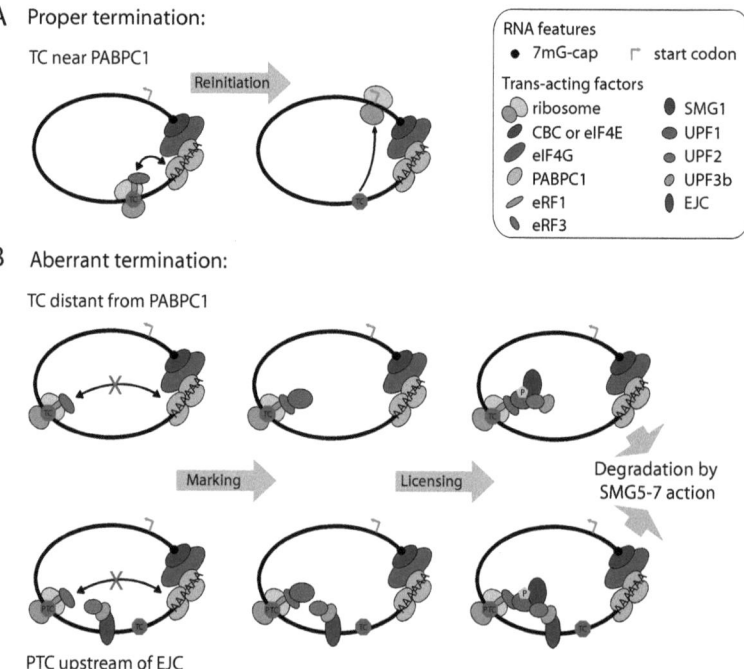

FIGURE 5: **The unified model for PTC recognition.** (A) Similar to the *faux* 3'UTR model, the unified model for PTC recognition proposes, that in the vicinity of the poly(A) tail, a PABPC1-mediated signal promotes proper translation termination, resulting in efficient reinitiation of the ribosome at the 5' end of the mRNA and in a stable mRNP. (B) If the ribosome terminates at a TC too far away from the poly(A) tail to receive the PABPC1-mediated translation-termination promoting signal, UPF1 can bind to the stalled ribosome instead, thereby marking this TC as premature. Subsequently, UPF2 and UPF3b interact with UPF1, promoting SMG1-mediated phosphorylation of UPF1. This licensing step commits the mRNA to rapid degradation by yet unknown pathways that involve SMG5-7 binding to the phosphorylated UPF1 (upper part). An EJC downstream of a TC can function as an NMD enhancer by shortening the time window between UPF1 binding and its phosphorylation by locally concentrating UPF2 and UPF3b (lower part). (adapted from paper IV)

2.3.5. Focusing on UPF1 function

Although NMD has been studied intensively during the last 30 years, we are far away from understanding the exact molecular events that lead from the recognition of the PTC to the degradation of the mRNA. Understanding the molecular mechanism of translation termination will be a central issue. And even more important, conceiving the function and regulation of UPF1 will bear the key to understand the mechanism of NMD at the molecular level. Therefore I chose two different approaches to gain more insight into UPF1 function: One approach to study the localization of UPF1 at different stages along the NMD pathway using a laser scanning microscope (LSM)(see paragraph 2.3.6.1.), and another approach to purify biochemically specific mRNPs stuck at certain stages in the NMD pathway to identify and quantify the interaction partners (see paragraph 2.3.6.2.).

2.3.5.1. Processing bodies are not required for mammalian NMD

After the step of PTC recognition (i.e. after the marking and licensing steps, see Fig. 5), phosphorylated UPF1 interacts with SMG5, SMG7 and/or SMG6, which finally leads to the degradation of the mRNA. But the mechanism, by which SMG5, SMG6 and SMG7 participate in the degradation, was largely unknown when I started this project. In human cells, SMG7 was found to co-localize with P-bodies, and UPF1 and SMG5 co-localize with P-bodies when SMG7 is overexpressed (Unterholzner & Izaurralde, 2004; Fukuhara et al., 2005; Durand et al., 2007). Furthermore, PTC+ mRNAs localize to P-bodies in a Upf1p-dependent manner in *S. cerevisiae*, and the ATPase activity of Upf1p is required at a later stage of NMD to trigger mRNA decapping and degradation (Sheth & Parker, 2006). Consistently, Upf1p mutants that are ATPase defective, but still allow RNA binding, accumulate large P-bodies (Cheng et al., 2007). In contrast, a Upf1p mutant with a defective RNA binding domain does not accumulate P-bodies. Interestingly, a combined Upf1p mutant with a defective RNA binding domain and a defective ATPase leads to the accumulation of P-bodies (Cheng et al., 2007). This suggests that the RNA binding is not required for P-body accumulation, and that Upf1p binds the RNA after ATP hydrolysis. Collectively, this data indicates that the localization of Upf1p-bound PTC+ mRNA to P-bodies represents an intermediate stage of NMD in *S. cerevisiae*, which can therefore be used to map activities of Upf1p that act upstream or downstream of P-body targeting. Furthermore, based on findings in mammalian cells that factors of the mRNA degradation machinery accumulate in P-bodies (reviewed in (Eulalio et al., 2007)), and that SMG5 and UPF1 localize to P-bodies in a SMG7-dependent manner, it was suggested that P-bodies might also play an important role in mammalian NMD. Therefore, I decided to investigate the role of P-bodies for mammalian NMD, with the ultimate goal to map specific activities of UPF1 upstream or downstream of the targeting to P-bodies.

To start this project, I received different UPF1 mutants from Akio Yamashita: UPF1 C126S (hereafter called CS; an analogous substitution in *S. cerevisiae* causes a loss of UPF2 binding)(Weng et al., 1996), K498Q (hereafter called KQ; abolishes the ATP binding capacity of UPF1, and therefore the ATPase is defective) (Cheng et al., 2007) and 4SA (four serins in UPF1 preferentially phosphorylated by SMG1 are mutated to alanins: S1073A, S1078A, S1096A and S1116A)(Kashima et al., 2006)(Fig. 6A). To stain the P-bodies, I received two different human reference sera ("IC6" and "18033") from Marvin Fritzler's lab (Ou et al., 2004; Bloch et al., 2006). Both sera contain a mixture of antibodies, whereas the IC6 serum mainly contains antibodies against Ge-1 (also known as

Hedls or EDC4) and the 18033 serum mainly antibodies against GW182 (also known as TNRC6A) (Fenger-Gron et al., 2005; Bloch et al., 2006). A typical picture of plain HeLa cells stained either with the IC6 serum or with the 18033 serum together with an α-DCP1a antibody is shown in Fig. 6B. The IC6 and 18033 sera both weakly stain the cytoplasm and strongly stain cytoplasmatic foci, which completely co-localize with the foci stained by the α-DCP1a antibody. This suggests that both sera stain P-bodies. Additionally, the IC6 serum also stains the nuclear lamina. Notably, a fraction of DCP1a was always observed in nuclear foci that might represent nucleoli (Fig 6B). Whether this represents unspecific binding of the antibody or whether some DCP1a indeed resides in the nucleoli remains unclear. Next, I transfected plasmids encoding the different N-terminally HA-tagged UPF1 mutants into HeLa cells, and 48 h later the cells were fixed, permebealized and stained with the antibodies. HA-UPF1 WT, HA-UPF1 CS and HA-UPF1 4SA were distributed quite evenly throughout the cytoplasm, whereas HA-UPF1 KQ accumu-

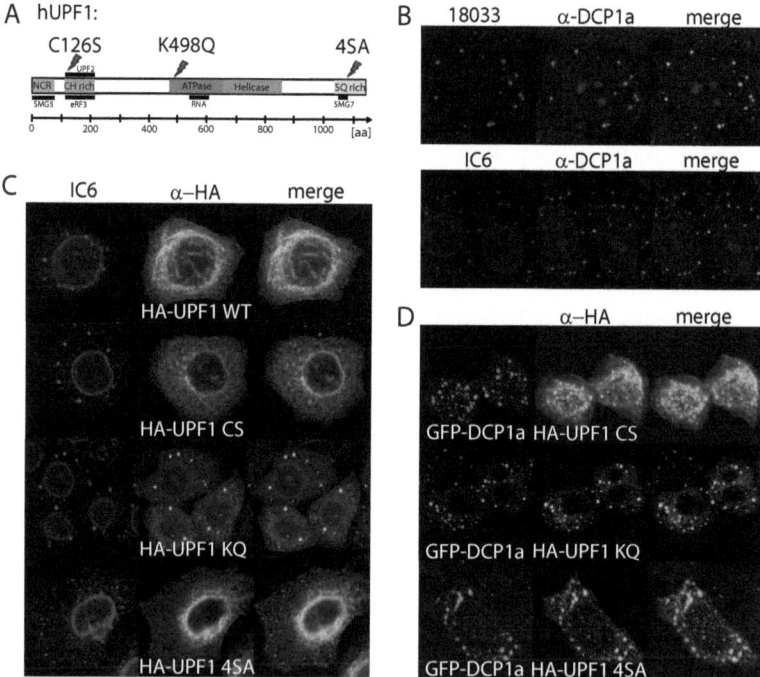

FIGURE 6: **Localization of UPF1 mutants.** (A) Schematic picture of human UPF1, the positions of the different mutations are indicated. The protein is drawn to scale with respect to the number of amino acids that they have, indicated by the scale bar at the bottom. Direct interactions with other proteins or RNA are shown as black bars. Reported interactions for which the interaction domains are not yet mapped are not indicated. hUPF1 contains an N-terminal conserved region (NCR), a CH-rich domain, a core domain comprising the ATPase and helicase, and an SQ-rich C-terminal domain. (B) HeLa cells stained with α-DCP1a antibody (green) together with the 18033 or IC6 serum (red), respectively. A merged picture is shown on the right side. (C) HeLa cells stained with the IC6 serum (red). HA-UPF1 WT or HA-UPF1 mutants were expressed from transfected plasmids and detected with α-HA antibody (yellow). (D) GFP-DCP1a (green) was co-expressed from a transfected plasmid together with the different UPF1 mutants (yellow).

lated in cytoplasmatic foci, which co-localize with the P-bodies stained by the IC6 serum (Fig. 6C). Notably, in cells transfected with HA-UPF1 4SA, the number of P-bodies per cell was increased, whereas the size of the P-bodies was decreased, compared to cells transfected with HA-UPF1 WT, HA-UPF1 KQ or HA-UPF1 CS, respectively. Whether this is a specific, functional consequence of the UPF1 4SA mutant remains elusive. Additionally, I co-transfected a plasmid encoding an eGFP-DCP1a fusion protein, together with the different UPF1 plasmids. Surprisingly, when eGFP-DCP1a was overexpressed, HA-UPF1 CS and HA-UPF1 4SA also accumulated in cytoplasmatic foci, which co-localized with the P-bodies (Fig. 6D). It has been shown that UPF1 interacts with DCP1a (Lykke-Andersen, 2002), and therefore the simplest interpretation of this discrepancy is that UPF1 is sequestered into P-bodies through the artificially high amounts of DCP1a. Importantly, HA-UPF1 K498Q was the only mutant which co-localized with P-bodies independent of high amounts of DCP1a, therefore suggesting a functional link between NMD and P-bodies.

Next, I wanted to test, if the accumulation of HA-UPF1 KQ in cytoplasmatic foci was dependent on the integrity of the P-bodies. Therefore, I knocked-down GW182, which has been shown to be critical for the integrity of P-bodies, and depletion of GW182 led to the disruption of the P-bodies (Yang et al., 2004; Lian et al., 2007). 96 h after the transfection of the shRNA encoding plasmids, the cells were fixed, permebealized and stained with the antibodies. Upon the knock-down of GW182, the mRNA level of GW182 was reduced to 25 % relative to the control knock-down (scr)(Fig. 7A, left panel). However, I was not able to monitor the knock-down efficiency on the protein level, because GW182 was not detectable by western-blot with the commercially available antibody (clone 4B6) (Fig. 7A, right panel). However, contradictory to previous publications, GW182 depleted cells mostly contained several microscopically detectable P-bodies (Fig. 7B). This suggests that either my knock-down conditions were not optimal, or the siRNAs did not target all isoforms of GW182 that are important for the formation of P-bodies. Nevertheless, I tested the effect of the GW182 knock-down on the mRNA levels of our Ig-µ NMD reporter. The mRNA levels of Ig-µ WT increased to about 265 % upon the knock-down of GW182 relative to the control knock-down (scr), and in a combined knock-down of GW182 with UPF1, the mRNA levels of Ig-µ WT increased to about 220 % compared to the control knock-down (Fig. 7C). The mRNA levels of Ig-µter310 were reduced to 2 % relative to the control knock-down of Ig-µ WT. The mRNA levels of Ig-µter310 increased 14-fold to 28 % upon the combined knock-down of UPF1 and GW182, but remained almost constant when only GW182 was knocked-down. This suggests that GW182 is not involved in mammalian NMD. However, this experiment did not allow any conclusions about a potential functional link between mammalian NMD and P-bodies. Because the GW182 knock-down did not have the expected effect on the integrity of P-bodies, and because I additionally had problems with the detection of GW182 by western-blot, I decided to knock-down Ge-1, another P-body marker protein.

Depletion of Ge-1 led to the disruption of microscopically detectable P-bodies, as judged by the uniform cytoplasmatic distribution of DCP1a under these conditions (see paper VI, Fig. 2A). The results from the subsequent experiments are part of paper VI, and therefore I will only shortly summarize the results. In this study, I could show that the exogenously expressed ATPase-defective UPF1 KQ mutant and a minor fraction of the endogenous UPF2 and UPF3b accumulate in cytoplasmic foci that co-localize with P-bodies in HeLa cells, whereas wild-type UPF1 was not detected in P-bodies. The ATPase-

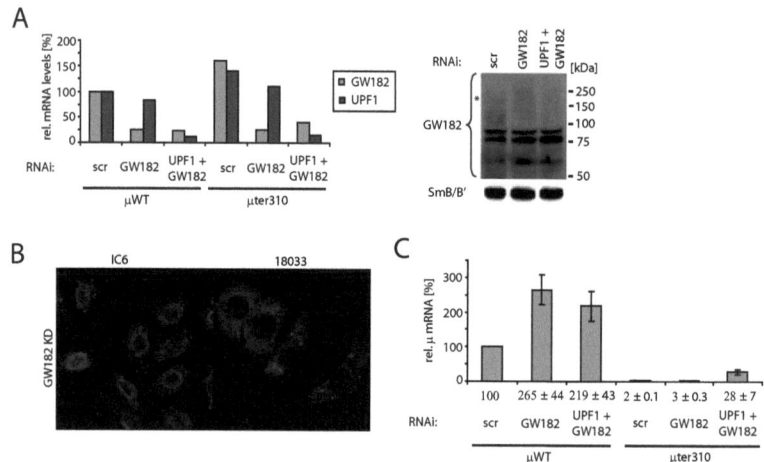

FIGURE 7: **RNAi mediated depletion of GW182.** (A) The efficieny of the GW182 knock-down or the combined knock-down of GW182 together with UPF1 was assessed by real-time RT-PCR (left panel) and by western-blot (right panel). The HeLa cells were harvested 96 h after the transfection of the shRNA encoding pSupuro plasmids. Real-time RT-PCR data from one run is shown. The asterisk indicates the expected position of GW182 on the western-blot, SmB/B' served as loading control. (B) Knock-down of GW182 performed as in (A). The HeLa cells were stained either with the IC6 serum or the 18033 serum. (C) The mRNA levels of the co-transfected NMD-reporter genes Ig-μWT or Ig-μter310 of the GW182 and UPF1 knock-down shown in (A) were measured by real-time RT-PCR. Average values from one experiment and 3 real-time RT-PCR runs are shown.

defective UPF1 co-localized with P-bodies independent of UPF2 and UPF3b, whereas the localization of the UPF1 mutant, UPF2 and UPF3b in cytoplasmic foci was dependent on the presence of microsopically detectable P-bodies. Strikingly, the disruption of the P-bodies by RNAi-mediated depletion of Ge-1 did no affect the mRNA levels of PTC+ reporter genes, and also endogenous NMD-substrates were not affected under these conditions. Collectively, these results demonstrate that the depletion of microscopically detectable P-bodies does not impair mammalian NMD. Nevertheless, these experiments do not exclude the possibility that some steps of NMD can occur in P-bodies under physiological conditions. However, I favour the interpretation that the P-body-localized NMD factors are not active in NMD, and therefore I suggest that P-bodies are not involved in mammalian NMD.

2.3.5.2. Purification of UPF1 bound PTC+ mRNPs

The biochemical properties and enzymatic activities of UPF1 have been studied intensively during the last 10 years. UPF1 interacts with the eukaryotic release factors eRF1 and eRF3 (Czaplinski et al., 1998; Wang et al., 2001; Kashima et al., 2006; Ivanov et al., 2008), it binds UPF2 through its CH-rich region (Kadlec et al., 2006), and it interacts with SMG1, SMG5, SMG6 and SMG7. Furthermore, UPF1 undergoes a cycle of phosphorylation and dephosphorylation, which is regulated by the other NMD factors. UPF1 exhibits RNA binding, RNA-dependent ATP hydrolysis, and 5'-to-3' ATP-dependent RNA helicase activities, and inhibition of any of these activities suppresses NMD (Bhattacharya et al., 2000; Kashima et al., 2006; Cheng et al., 2007). Yet, the protein-interaction network and the regulation of the different enzymatic activities of UPF1 at the different stages of NMD are not clear. Therefore I wanted to shed more light onto the interactions and the regulation of UPF1 at different stages of NMD in an unbiased way. NMD is a highly dynamic process, which makes the observation of transient intermediate stages of NMD difficult. To deal with this problem, NMD needs to be arrested at certain stages along the pathway, i.e. by the overexpression of UPF1 mutants, thereby leading to the accumulation of these mRNPs. The challenge of this project is the specific purification and analysis of those mRNPs. I decided to start with the same HA-tagged UPF1 mutant I used in the localization studies (see 2.3.5.1) to set up a two-step purification protocol. In the first purification step, the co-expressed NMD reporter, containing MS2-binding sites in the 3'UTR, can be purified over an amylose-coupled agarose column using a fusion-protein of maltose-binding proten (MBP) with the MS2-binding protein (MS2MBP)(Fig. 8A). An advantage of this system is that the elution step can occur under mild conditions (i.e. with 150 mM NaCl and at 4 °C), where the mRNP should remain native. The second purification step is then an affinity purification with an α-HA antibody, to enrich those mRNPs bound by the HA-tagged UPF1 mutant (Fig. 8A). Finally, the proteins can be identified by mass spectrometry, and if necessary quantified by the use of SILAC mass spectrometry. This approach should allow the specific enrichment of PTC+ mRNPs stuck in the NMD pathway by the bound UPF1 mutant, and therefore the analysis of the protein composition should reveal which proteins are involved at this specific stage of NMD. This approach is very ambitious and several problems had to be solved. According to Henning Urlaub (Göttingen, Germany), who had agreed to perform the mass-spectrometry analysis, the minimal amount needed to perform the mass spectrometry analysis was 1 pmol of purified mRNPs. Assuming an expression of the NMD reporter in a range usually observed in our HeLa cells, an estimate indicated that 10^{10} HeLa cells would be necessary to generate 10 pmol of mRNPs (data not shown). This high amount of cells is not feasible for transient transfections, therefore HeLa cells stably expressing the NMD reporter and the UPF1 mutant had to be generated. Additionally, to avoid toxic effects of the UPF1 mutants during the growth phase of the cells, the Tet-On system was used to make the expression of the UPF1 mutant inducible upon the addition of doxycyline. First, HeLa polyclonal cell pools were generated expressing the Tet-repressor protein (Fig. 8B), resulting in the HeLa wi TR cell line. To test this cell line, UPF1 KQ was transfected transiently, and then the cells were cultivated for 48 h in in medium containing 1 µg/ml doxycyline. As shown by the western-blot in Fig. 8C, the expression of UPF1 KQ in the HeLa wi TR cell line could be induced to almost similar levels as in the control HeLa wi cells, which do not express the Tet-repressor protein. Next, a plasmid encoding the tet-inducible MS2MBP fusion-protein was stably integrated into the HeLa wi TR cell line, resulting in the HeLa wi TR MS2MBP cell line (Fig. 8B). Into this cell line, the plasmid encoding the NMD reporter gene with the MS2 binding sites in the 3'UTR was stably integrated. The Ig-µ minigene, the TCRβ minigene and the β-globin minigene were used as NMD reporters, thus resulting in the cell lines HeLa wi TR MS2MBP µter310, HeLa wi TR MS2MBP TCRβ or HeLa wi TR MS2MBP βGl. A first trial of the MS2MBP purification step was done using the HeLa wi TR MS2MBP βGl cell line. Judged from a silver-stained SDS-page, a few specific bands could be enriched (data not shown), and it was possible to measure an enrichment of βGl mRNA in the eluate (data not shown). Thus, the MS2MBP purification step worked in my hands, and finally the N-terminally HA-tagged UPF1 mutants were stably integrated into the cells expressing the Tet-repressor, the MS2MBP fusion-protein

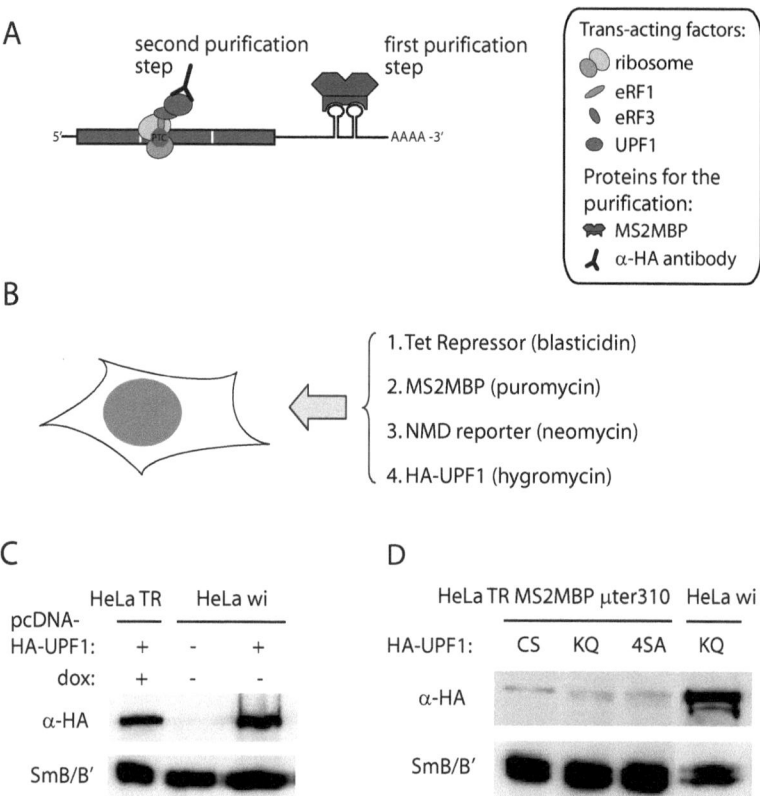

FIGURE 8: **Double-step purification of UPF1-bound mRNPs.** (A) Schematic depiction of the double-step mRNP purification strategy. (B) Schematic depiction of the different transfection steps to generate the stable HeLa cell lines for the mRNP purification. (C) Western blot to assess the expression level of HA-UPF1 KQ from the Tet-inducible pcDNA3.1 plasmid in the HeLa cell line expressing the Tet-repressor protein (HeLa TR). HeLa TR and plain HeLa cells (HeLa wi) were transiently transfected with pcDNA3.1 HA-UPF1 KQ. The medium for the HeLa TR cells contained 1 µg/ml dox, and the cells were harvested 72 h after the transfection. Detection of SmB/B' served as loading control. (D) Western blot of HeLa cells stably expressing the different UPF1 mutants. HeLa TR MS2MBP µter310 cells stably expressing HA-UPF1 CS, HA-UPF1 KQ or HA-UPF1 4SA, respectively, were cultivated in medium containing 1 µg/ml dox for 48 h. Transiently expressed UPF1 KQ (same sample as lane 3 in (C)) served as expression control, detection of SmB/B' served as loading control.

and an NMD reporter gene (Fig. 8B). The expression of the UPF1 mutants was tested in the HeLa wi TR MS2MBP µter310 cell line. But, due to unknown reasons, the expression of the UPF1 mutants in the stable cell lines was very low compared to the transiently expressed UPF1 KQ in the HeLa wi cells, as judged from the western blot in Fig. 8D. It was therefore not surprising that in a first trial of the double-step purification, no proteins and no RNA was detectable in the eluates (data not shown). To deal with this expression problem, I tried to integrate the UPF1 mutant by lentiviral transduction, which was supposed to result in a robust and strong expression of the transduced protein of interest.

Unfortunately, several trials to produce stable cell lines by lentiviral transduction failed. The simplest explanation for this failure might be that UPF1 has a size around the upper limit for the lentiviral system (UPF1 is 1118 amino acids long). In parallel, the same set of stable cell lines was established with HeLa spinner cells. But similar to the adherent HeLa cells, the expression of the stably integrated UPF1 mutants was very low (data not shown).

The purification of high amounts of UPF1-bound PTC+ mRNPs was technically not feasible, and therefore we decided to simplify the purification process (one step pulling at the RNA, one step pulling at UPF1) by the use of a RNA tandem affinity tag (RAT, Fig. 9A; see also paragraph 3.2) (Hogg & Collins, 2007). The purification of RAT-tagged reporter genes would allow the comparison of the proteins bound to PTC- or PTC+ mRNPs. The RAT tag consists of a binding site for the pseudomonas aeruginosa phage 7 (PP7; first purification step) and a tobramycin-binding aptamer (second purification step). To test this purification method, I cloned the RAT tag into the 3'UTR of the βGl reporter gene (βGl RAT), and a βGl reporter gene with a RAT tag inserted in the reversed orientation (βGl rev) served as control. To monitor the binding of the βGl RAT RNA during the purification process, I decided to use a radioactive ^{32}P-labelled RNA. A PCR-product of βGl RAT or βGl rev was *in vitro* transcribed by T7 RNA polymerase (Fig. 9A, left panel), and then the labeled βGl RAT or βGl rev RNA was added to a lysate of plain HeLa cells. As shown in Fig. 9B (right panel), βGl RAT RNA was present in the eluate (position indicated by the arrow), whereas the βGl rev RNA was remaining in the supernatant of the first purification step and no signal was detectable in the eluate of the control purification with βGl rev. The fraction of RNA with the lower mobility may represent protein-bound RNAs (i.e. with the PP7 protein). I started with this purification only recently and will

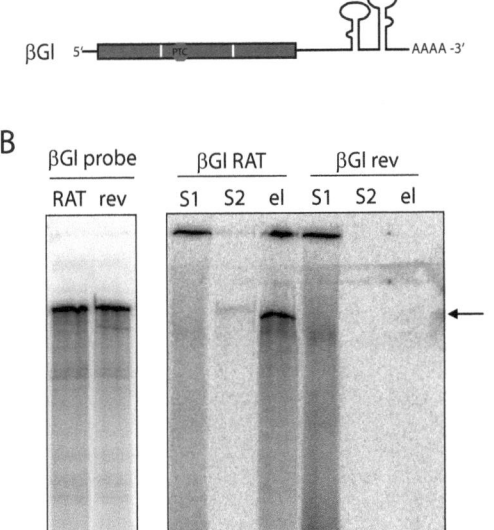

FIGURE 9: **Double-step purification of mRNPs using the RAT tag.** (A) Schematic depiction of the βGl reporter containing the RAT tag in the 3'UTR. (B) The radioactive labeled in vitro transcript of the RAT-tagged βGl reporter was separated on a 5 % denaturing page (left panel). A βGl reporter with the RAT tag inserted in the other direction served as control (rev). The *in vitro* transcript of the RAT-tagged βGl reporter was added to a lysate of plain HeLa cells and RAT purified. S1 represents the supernatant of the first purification step, before the elution with the TEV protease. S2 represents the supernatant of the second purification step, before the elution of the tobramycin beads. El represents the eluate from the tobramycin beads. The samples were separated on a 5 % denaturing page, the expected position of the transcript is indicated by the arrow.

proceed with this project in the future.

2.3.6. Involvement of NMD factors in other cellular processes

During the last 10 years, investigations have shown that the NMD factors are involved in various other cellular processes (reviewed in paper V, and (Isken & Maquat, 2008)). For instance, UPF1, UPF2 and UPF3b, and the EJC factors Y14, MAGOH and RNPS1 have been reported to stimulate translation (Nott et al., 2004). In *S. cerevisiae*, NMD factors were shown to increase translation termination fidelity, and null-mutant strains of Upf1p, Upf2p or Upf3p exhibit nonsense suppression (Maderazo et al., 2000; Wang et al., 2001). A recent report by Ivanov and colleagues shows that UPF1 decreases the translation termination fidelity in human cells (Ivanov et al., 2008), indicating a fundamental difference in the function of the yeast Upf1p and the human UPF1 proteins in the mechanism of translation termination.

Interestingly, NMD factors are involved in telomere maintenance and genome stability. SMG5 and SMG6 interact with telomerase, and UPF1, UPF2, SMG1 and SMG6 were all found to be enriched at telomeres, and they negatively regulate the association of the telomeric repeat-containing RNAs (TERRAs) with chromatin (Reichenbach et al., 2003; Snow et al., 2003; Azzalin et al., 2007). Furthermore, depletion of UPF1, SMG1 and SMG6 generates chromosome and chromatid breaks, which may explain the cell-cycle arrest in early S phase upon depletion of UPF1 (Azzalin & Lingner, 2006; Azzalin et al., 2007). Moreover, SMG1 phophorylates p53 and other substrates involved in checkpoint signaling, and UPF1 is phosphorylated by the PI-3 kinase-related kinases (PIKKs) ATM (ataxia telangiectasia mutated), ATR (ataxia-telangiectasia mutated and Rad 3-related), and DNA-PK (DNA-dependent protein kinase) as a result of DNA damage or DNA replication block responses (Brumbaugh et al., 2004; Kaygun & Marzluff, 2005; Muller et al., 2007). Thus, it is suggested that UPF1 phosphorylation by different PIKKs can variously influence UPF1 function.

In addition, studies in *C. elegans* revealed that SMG-2 (*C. elegans* UPF1), SMG-5, and SMG-6 are required for establishing persistent RNAi-mediated gene silencing. Mutant SMG-2, SMG-5, and SMG-6 animals recovered within 4 days from *unc-54* RNAi-mediated paralysis, whereas wildtype worms did not (Domeier et al., 2000). Consistently, a functional link between UPF1 and RNAi has recently been reported for plants (Arciga-Reyes et al., 2006). In contrast, UPF1 is not involved in RNAi in *D. melanogaster* (Kim et al., 2005; Rehwinkel et al., 2005). It remains to be shown, if UPF1 also functions in the mammalian RNAi pathway, as recently a physical link between NMD and RNAi factors has been reported: UPF1 co-immunoprecipitates AGO1 and AGO2 in an RNAse A insensitive manner, and UPF1 was identified in AGO1-associated mRNPs (Hock et al., 2007).

2.3.7. Therapeutic aspects

It has been estimated that ~30 % of the known disease-associated mutations generate a PTC, which suggests that NMD has a widespread impact on the phenotype of numerous genetic diseases (Holbrook et al., 2004). NMD is beneficial if it prevents the production of C-terminally truncated proteins that would have had dominant-negative effects. In such cases, PTCs that efficiently trigger NMD lead to a mild, recessive phenotype, whereas PTCs that are recognized inefficiently lead to a severe phenotype. The beneficial role of

NMD was first described on β-thalassemic diseases, where the production of β-globin and therefore of hemoglobin is impaired by a PTC in the β-globin ORF (Chang & Kan, 1979). In the common recessive form of β-thalassemia, the faulty transcripts are degraded by NMD; hence heterozygotes express enough functional hemoglobin for a normal phenotype, whereas homozygotes or patients with PTCs that are inefficiently recognized by NMD (i.e. PTCs in the last exon) suffer from a severe anemia (Thein et al., 1990; Kugler et al., 1995). By contrast, NMD is detrimental if it prevents the production of truncated proteins that still have residual function, as has been described for frequent mutations causing cystic fibrosis or duchenne muscular dystrophy (DMD). PTCs in regions of the dystrophin gene where they efficiently trigger NMD are associated with the severe form of DMD, while the milder form, the so-called Becker muscular dystrophy (BMD), is associated with nonsense mRNAs that are not recognized by NMD and hence serve as templates for the synthesis of C-terminally truncated but still functional dystrophin protein (Kerr et al., 2001). Yet, treatments for such diseases are not available, and because NMD plays an important role in the clinical manifestation of such diseases, the ability to specifically interfere with NMD will be crucial in the development of therapeutic treatments.

The use of NMD-modulating drugs has been tested for many years (Holbrook et al., 2004). Promising results were accomplished by the use of aminoglycoside antibiotics, which bind to the decoding center of the ribosome and decrease the accuracy of the codon-anticodon pairing, thereby promoting the readthrough of stop codons (also called nonsense suppression) (Kellermayer, 2006; Hermann, 2007). The aminoglycoside gentamicin suppresses stop codons in *in vitro* assays, and beneficial effects of gentamicin treatments were reported in clinical trials with patients suffering from nonsense-mediated cystic fibrosis (Wilschanski et al., 2003). But, such unspecific drugs also have various side-effects. On the one hand, the readthrough at normal stop codons will be promoted, leading to a change in global gene expression and to the production of C-terminally extended proteins (Welch et al., 2007). On the other hand, because NMD is involved in many processes of gene expression, such generally NMD-inhibiting drugs can be toxic for the cell. Recently, a new readthrough-promoting drug called PTC124 with no structural relationship to aminoglycosides has been described. Strikinlgy, PTC124 selectively induces readthrough of premature but not normal termination codons (Welch et al., 2007). PTC124 rescued striated muscle function in mdx mice expressing dystrophin nonsense alleles and the drug is currently being tested in clinical phase II trials on patients suffering from cystic fibrosis or DMD.

2.3.7.1. Suppression of NMD by an anti-sense oligo-nucleotide

In addition to mutations in the ORF, leading to diseases as described above, our unified NMD model predicts the existence of populations of NMD targets that have not been appreciated previously. On the one hand, many PTCs in the last exon have the potential to elicit NMD, especially when the last exon is long. On the other hand, various mutations in the 3'UTR have the potential to alter the spatial relationship between the TC and the poly(A) tail: insertions in the 3'UTR; mutations of the natural stop codon; mutations that destroy poly(A) sites or create cryptic ones; and modification of binding sites for RNA binding proteins. Any of these types of mutations could turn an mRNA into an NMD target. Treatments with readthrough-promoting NMD inhibitors like PTC124 are problematic in such cases, because they do not stabilize the mRNA and they would C-terminally extend the wildtype protein. In contrast, NMD suppression by our foldback strategy (see paper III) would augment wild-type protein levels and therefore represents a putative therapeutic approach, provided the foldback can be induced in *trans*. Importantly, this approach is sequence specific, and therefore it would not be expected to have toxic side-effects for the cell. For a proof-of-principle experiment, I used a 2'-O-methylated oligonucleotide that base pairs with its 5' half near the poly(A) tail and with its 3' half near the PTC of the Ig-μter310 reporter (Fig. 10A). Transfection of this double-target antisense oligonucleotide increased in a dose-dependent manner the NMD reporter mRNA level (Fig. 10B), demonstrating the feasibility of this approach. Transfection of the two halves of the double-target oligonucleotide as separate oligonucleotides served as a control.

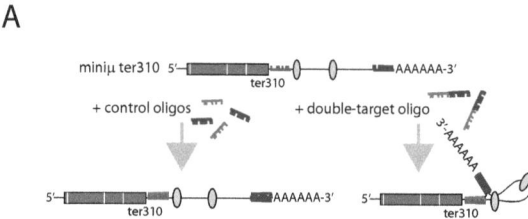

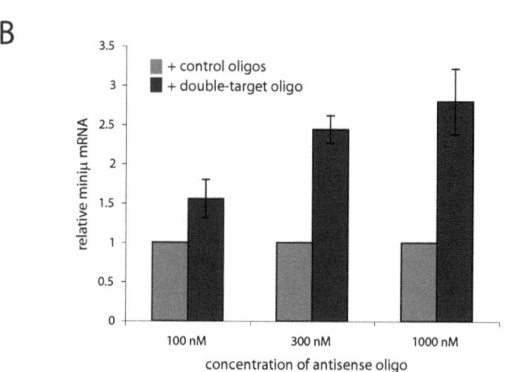

FIGURE 10: **Suppression of NMD in *trans* by folding the poly(A) tail near the PTC with help of a double-target antisense oligonucleotide.** (A) Schematic illustration of the NMD reporter Ig-μter310 and the co-transfected double-target 2'-O-methyl antisense oligonucleotide (right) or the two separated halves of the double-target oligo as a control (left). The miniμ sequences complementary to the 5' half (blue) and the 3' half (green) of the double-target oligo are depicted in purple and red, respectively. (B) Relative Ig-μter310 mRNA levels, normalized to β-globin WT mRNA from a cotransfected plasmid were determined. Five hours after transfection of the plasmids, 100, 300, or 1000 nM double-target or control oligos were transfected, and RNA was isolated 24 h thereafter. The mRNA level in the control sample was set 1 for each oligonucleotide concentration. Average values and SD from four real-time RT-PCR runs are shown.

This strategy might for example be promising in the case of a 3 kb retrotransposon insertion into the 3' UTR of the fukutin gene, which is found in 87% of the mutated alleles of Japanese patients suffering from Fukuyama-type congenital muscular dystrophy (FCMD) and which results in very low levels of fukutin mRNA (Kobayashi et al., 1998). Yet, the reason why the insertion of the retrotransposon into the 3'UTR of the fukutin gene leads to decreased mRNA levels is unknown, but we hypothesized that the mutated fukutin mRNA was turned into an NMD substrate by the insertion of the retrotransposon. If the fukutin FCMD mRNA with the retrotransposon insertion indeed is an NMD substrate, it was suggested to be the perfect target to test our oligonucleotide based approach on a disease-relevant mRNA. Unfortunately, Mr. Toda, the head of the Japanese group that published the paper by Kobayashi and colleagues in 1998, refused any collaboration with us, and as well he refused to send us the reporter plasmids of the fukutin gene used in their publication. Nevertheless, we were eager to test our hypothesis, and thus started to generate our own fukutin reporter constructs.

The fukutin gene spans a genomic DNA region of more than 100 kb and consists of 10 exons (Kobayashi et al., 2001). Exon 1 and a part of exon 2 contain the 5'UTR, the ATG initiation codon is in exon 2, and the stop codon is at the beginning of exon 10 (Fig. 11A). The exon 10 contains 3 AATAAA poly(A) sites at positions 2604, 2840 and 4769 bp downstream of the stop codon. Notably, the 3'UTR of the fukutin mRNA has a length of ~5 kb, which may suggest that this mRNA is already an NMD substrate. RT-PCR analysis on cDNA derived from plain HeLa cells revealed that the fukutin gene is expressed in our HeLa cells, and that the fukutin mRNA is alternatively spliced (Fig. 11B, left panel). Consistent with a previous publication (Kobayashi et al., 2001), the longest splice variant is the most abundant splice form (Fig. 11B, left panel). Furthermore, primers upstream of the distal poly(A) site (4769 bp downstream of the stop codon) gave a RT-PCR signal, suggesting that our HeLa cells express a fukutin mRNA with a 5 kb long 3'UTR (Fig. 11B, right panel), although this does not exclude that the other poly(A) sites are used as well. Importantly, the endogenous fukutin mRNA levels were not significantly stabilized in HeLa cells treated with cycloheximide (Fig. 11C, upper panel), or depleted of UPF1, UPF2 or UPF3b, compared to the control knock-down (Fig. 11C, lower panel).

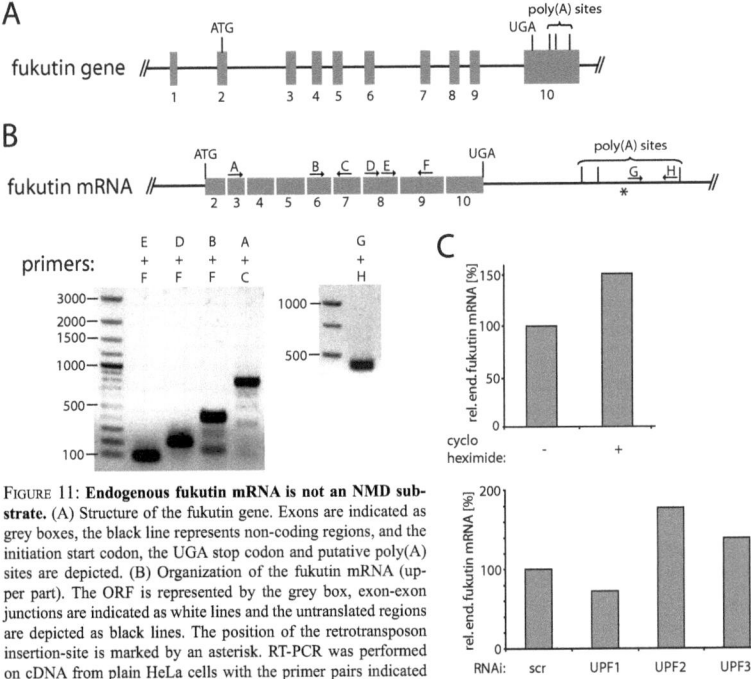

FIGURE 11: **Endogenous fukutin mRNA is not an NMD substrate.** (A) Structure of the fukutin gene. Exons are indicated as grey boxes, the black line represents non-coding regions, and the initiation start codon, the UGA stop codon and putative poly(A) sites are depicted. (B) Organization of the fukutin mRNA (upper part). The ORF is represented by the grey box, exon-exon junctions are indicated as white lines and the untranslated regions are depicted as black lines. The position of the retrotransposon insertion-site is marked by an asterisk. RT-PCR was performed on cDNA from plain HeLa cells with the primer pairs indicated (lower part). (C) Plain HeLa cells were treated with cycloheximide, and the cells were harvested 4 h later (left panel). RNAi-mediated depletion of UPF1, UPF2 and UPF3b on plain HeLa cells (right panel). Values from one real-time RT-PCR run are shown.

A
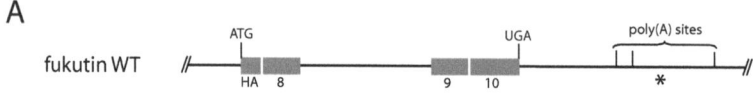

fukutin WT

B
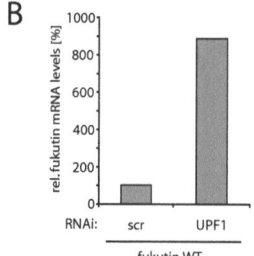

FIGURE 12: **The fukutin reporter gene does not behave like the endogenous fukutin mRNA.** (A) Structure of the fukutin WT and reporter gene. The exons are represented by grey boxes, exon-exon junctions are indicated as white lines and the untranslated regions are depicted as black lines. The position of the retrotransposon insertion-site is marked by an asterisk. (B) RNAi-mediated depletion of UPF1 in HeLa cells. The fukutin WT or the fukutin hyb1 reporter plasmids were co-transfected together with a plasmid encoding a β-globin WT gene for normalization. 96 h after the transfection of the plasmids, the cells were harvested and analyzed by real-time RT-PCR. Relative values from one real-time RT-PCR run are shown.

This demonstrates that the endogenously expressed fukutin mRNA in the HeLa cells is not an NMD substrate.

With these results in mind, I started the cloning of a fukutin minigene. The region spanning exons 8 and 9 with the intron in between and the exon 10 were amplified out of genomic DNA and ligated into a pcDNA3.1-HA vector, resulting in the fukutin WT construct (Fig. 12A). Next, I wanted to insert the 3 kb retrotransposon into the 3' UTR of the fukutin WT minigene, to mimic the FCMD mRNA. We received genomic DNA from a Japanese homozygous FCMD patient, and I tried to amplify the retrotransposon. Unfortunately, any effort to PCR-amplify the retrotransposon failed, most probably because the insertion is highly repetitive and GC-rich, and after two month of doing PCR, we decided to let synthesize the retrotransposon by a company. But before doing so, I wanted to test the cloned fukutin WT minigene. Surprisingly, the expression of the transiently transfected fukutin WT reporter was (with Ct-values around 37 in the real-time RT-PCR assay) very low, compared to the endogenously expressed fukutin (Ct-values around 27; data not shown). Upon RNAi-mediated depletion of UPF1, the fukutin WT mRNA levels increased about 8-fold, compared to the control knock-down (Fig. 12B). Thus, due to unknown reasons, the fukutin WT reporter gene is an NMD substrate and is very low expressed. The fukutin reporter gene is correctly spliced, fully sequenced and also the vector is correct (data not shown), but until today, I could not find a reason for the differential behaviour of the fukutin minigene compared to the endogenous fukutin mRNA. This issue will be subject to further investigation.

2.3.7.2. A combinatorial approach of PTC124 together with the double-target anti-sense oligonucleotide

Another advantage of using a double-target anti-sense oligonucleotide to stabilize a PTC+ transcript is its sequence specifity, and therefore it has the potential to specifically stabilize any NMD substrate. It has been shown that PTC124 does not affect the overall gene expression levels, and also no stabilizing effect was found on mRNAs bearing a PTC (Welch et al., 2007). Therefore, a combined treatment with PTC124 and a double-target anti-sense oligonucleotide may lead to a higher production of full-length protein than could be achieved with PTC124 alone. The stabilization of the PTC+ mRNA by the double-target anti-sense oligonucleotide also would lead to the production of truncated proteins, and therefore this combined approach might only be beneficial for diseases, where the truncated protein does not exhibit dominant-negative effects. To test this hypothesis on our HeLa cells, we started a collaboration with the group of C. Leumann (Department of Chemistry and Biochemistry, Bern). M. Stoop then synthesized the small organic molecule PTC124 chemically. However, in first tests, I was not able to detect readthrough promotion of PTC124 on our Ig-µter440 reporter gene (data not shown). The termination codon 440 is an UAA, and PTC124 exhibited the lowest readthrough efficiency on UAA codons (Welch et al., 2007). Therefore I cloned a reporter plasmid containing a luciferase ORF with a UGA PTC at amino acid position 309, because UGA codons showed the highest readthrough efficiency in PTC124-treated cells (Welch et al. 2007), and a luciferase assay is much more sensitive than a western blot. Therefore, with this luciferase reporter plasmid, it should be possible to test PTC124 and the effect of a combination with a double-target anti-sense oligonucleotide on the production of full-length protein.

2.4. Nonsense-mediated transcriptional gene silencing (NMTGS)

Another mechanism to downregulate the expression of PTC-bearing mRNAs was recently discovered in our lab (Buhler et al., 2005). When Ig-µ or Ig-γ minigenes were stably integrated into the genome of the HeLa cells, the PTC-containing minigenes were transcriptionally silenced, and we termed this phenomenon nonsense-mediated transcriptional gene silencing (NMTGS). Intriguingly, I did not find reduced polymerase II occupancy on a silenced Ig-µ minigene (Fig. 13), but it remains elusive, whether this is mechanistically meaningful (see paragraph 3.5.). Among the six NMD reporter genes tested so far, only the Ig-µ and the Ig-γ minigenes showed NMTGS, indicating that NMTGS depends on a gene-specific signal that might be confined to immunoglobulin heavy-chain encoding genes. NMTGS is the result of chromatin remodeling from the active state, where H3K9 is acetylated, to the silent state, which is accompanied by H3K9 methylation (Buhler et al., 2005). More precisely, I could show that in the silenced state, the PTC-containing gene is more associated with mono-, di- and tri-methylated H3K9 (paper II). Consistently, NMTGS can be reversed by treating the cells with histone deacetylase (HDAC) inhibitors like sodium butyrate (SB) or trichostatin A (TSA) ((Buhler et al., 2005) and paper II). Furthermore, I could demonstrate that NMTGS requires the translation of its cognate mRNA, as it is not observed under conditions where translation of the PTC-containing mRNA is inhibited through an iron-responsive element (IRE). The IRE consists out of a stem loop located in the 5'UTR, which binds the iron regulatory proteins (IRPs) and

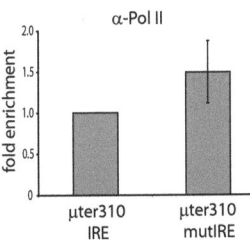

FIGURE 13: **Polymerase II occupancy is not reduced on a silenced Ig-µ minigene.** ChIP assays were performed with an antibody against polymerase II. The relative amount of Ig-µ and endogenous hGAPDH DNA in the immunoprecipitated fraction and in the fraction before immunoprecipitaition was quantified by real-time RT-PCR. The relative enrichment of immunoprecipitated Ig-µ DNA, normalized to the enrichment of the endogenous hGAPDH DNA is represented as fold enrichment. Average values from 4 independent ChIP experiments are shown.

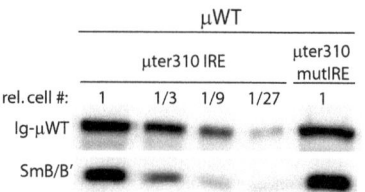

FIGURE 14: **NMTGS can not be induced in *trans*.** A polyclonal HeLa cell pool stably expressing Ig-µWT was stably transfected either with Ig-µter310 IRE or Ig-µter310 mutIRE. A western blot with a serial dilution of cells expressing Ig-µWT and Ig-µter310 IRE compared to cells expressing Ig-µWT and Ig-µter310 mutIRE is shown. Detection of SmB/B' served as loading control.

thereby inhibits the translation of this mRNA (Klausner et al., 1993; Hentze & Kuhn, 1996). A mutant IRE (mutIRE) that does not bind IRP served as control. The coupling of a translation-dependent event, the recognition of the PTC, with the transcriptional silencing suggests the existence of a feedback signal, which may act in *trans*. However, the detection of a putative Ig-µ specific siRNA by northern blot or by RNase protection assay failed so far (O. Mühlemann and R. Gudipati, unpublished observations). Therefore I decided first to test, if NMTGS could be induced in *trans*. I stably integrated an Ig-µter310 IRE or an Ig-µter310 mutIRE reporter gene into a HeLa cell line, which already stably expressed an Ig-µWT reporter gene. The expression of the Ig-µter310 mutIRE would then be expected to induce NMTGS and by doing so, to silence the already integrated Ig-µWT reporter gene, given that NMTGS can be elicited in *trans*. The integration of the Ig-µter310 IRE reporter gene served as control, as its translation is inhibited by the IRE. However, as judged by the western blot in Fig. 14, the expression of the Ig-µter310 mutIRE reporter gene did not reduce the amount of Ig-µWT protein, suggesting that NMTGS can not be induced in *trans*.

NMD and NMTGS are both triggered by the recognition of a PTC, suggesting that the mechanisms of NMD and NMTGS might be at least partially coupled. RNAi-mediated depletion of UPF1 reversed NMTGS, which suggests a mechanistic link between NMD and NMTGS (paper II). So far, we could demonstrate that NMTGS is PTC-specific; we have observed transcriptional silencing only with PTC+ Ig-µ and Ig-γ minigenes but not with missense or silent mutations in these genes (Buhler et al., 2005). Importantly, the results that translation and UPF1 are required for NMTGS provide a direct link between the transcriptional silencing and the machinery of PTC recognition.

3. DISCUSSION AND PERSPECTIVES

3.1. The mechanism of NMD is more conserved than previously appreciated

During the last five years, several studies performed in *S. cerevisiae, C. elegans, D. melanogaster, A. thaliana* and humans have been published, which provide evidence for an evolutionarily conserved mechanism of PTC recognition (Amrani et al., 2004; Buhler et al., 2006; Behm-Ansmant et al., 2007; Longman et al., 2007; Ivanov et al., 2008; Silva et al., 2008; Singh et al., 2008)(and paper III)(in detail described in paper IV). Based on these reports, we propose a "unified" model, which essentially extends the *faux* 3'UTR model to all eukaryotes studied so far. However, several lines of evidence suggest that the steps after PTC recognition diverge among eukaryotes and that a major evolutionary step occurred between *S. cerevisiae* and higher metazoans. First, homologs of the NMD factors SMG1-7 have not been identified in all eukaryotes (see table 2 in paper V), which already indicates that the mechanism of NMD can not be completely conserved. Second, based on our unified model for PTC recognition, it is suggested that the NMD pathway for *S. cerevisiae* diverges from the pathway of metazoans at the licensing step (see Fig 5). Finally, decapping is a major consequence of PTC recognition in *S. cerevisiae* (Muhlrad & Parker, 1994), whereas in *D. melanogaster* PTC recognition leads to endonucleolytic cleavage of the mRNA (Gatfield & Izaurralde, 2004). Two papers provide evidence that PTC recognition can also lead to accelerated deadenylation in *S. cerevisiae* and mammals (Cao & Parker, 2003; Chen & Shyu, 2003). However, recent results from our lab (Eberle et al., under revision) show that PTC recognition in mammals mainly leads to SMG6-mediated endonucleolytic cleavage of the mRNA, suggesting that the major consequence of PTC recognition, the endonucleolytic cleavage of the PTC+ mRNA, may be conserved in metazoans. Consistently, RNAi-mediated depletion of the decapping enzyme DCP2 (Eberle et al., under revision) and the decapping co-activator Ge-1 (paper VI) do not suppress NMD. Moreover, PTC+ mRNAs are targeted and degraded in a Upf1p-dependent manner in P-bodies in *S. cerevisiae* (Sheth & Parker, 2006), whereas in *D. melanogaster* and humans, P-bodies seem not to be required for NMD (Eulalio et al., 2007) (and paper VI).

Contrary to the downstream marker model, PTCs can trigger NMD in the absence of a downstream EJC or DSE (paper III) (Amrani et al., 2004; Buhler et al., 2006; Matsuda et al., 2007; Silva et al., 2008; Singh et al., 2008). However, it is apparent that the extent of mRNA downregulation in these examples of EJC-independent NMD is lower than in corresponding examples of NMD on transcripts with EJCs in the 3'UTR. Consistent with the idea that EJCs have an important role in NMD, knockdown of EJC core-factors in mammalian cells reduced the downregulation of many NMD reporter mRNAs (Gehring et al., 2003; Palacios et al., 2004; Shibuya et al., 2004; Buhler et al., 2006). In the light of this, we propose in the unified model that the EJC in mammals has evolved as a specialized second signal to enhance mammalian NMD. We hypothesize that, in mammals, under the evolutionary pressure to efficiently recognize and eliminate the large number of nonsense mRNAs produced by extensive alternative pre-mRNA splicing, the EJC as a spatial mark of previous splicing events has been incorporated into the mechanism of PTC recognition as an enhancer. Consistent with this view, proteins homologous to mammalian EJC core components are not involved in NMD in *D. melanogaster* and *C. elegans* (Gatfield et al., 2003; Longman et al., 2007), in which only a minor fraction of the pre-mRNAs is alternatively spliced. Notably, downstream sequence elements (DSEs) identified in *S. cerevisiae* might have NMD-enhancing functions similar to that of the EJC, by providing a binding platform for NMD enhancing factors (Gonzalez et al., 2000)(and papers IV and V).

3.2. Elucidating the function of UPF1

The importance of the various interactions and the enzymatic activities of UPF1 have been documented *in vitro* and *in vivo*, but not much is known about their function and regulation in the NMD pathway. The interpretation of data generated by RNAi-mediated knock-downs or overexpression of mutants is limited in this case, because UPF1 is also involved in several other cellular processes such as telomere maintenance, genome stability and translation (see paper V, and paragraph 2.3.6.). It was therefore exciting, when the work of Sheth and colleagues and the work of Cheng and colleagues were published (Sheth & Parker, 2006; Cheng et al., 2007), demonstrating that NMD and P-bodies are functionally linked in *S. cerevisiae*, and that the accumulation of P-bodies can serve as a mark of an intermediate stage in the pathway of NMD. Based on these studies, I decided to try a similar approach in HeLa cells. After I started this project by knocking-down GW182 and by studying the localization of the different UPF1 mutants received from Akio Yamashita, a paper by Eulalio and colleagues was published (Eulalio et al., 2007). They demonstrated that knocking down GW182 or Ge-1, which disrupted the visual P-bodies, did not affect the steady-state mRNA levels of an NMD reporter in *drosophila S2* cells. Consistently, I found that in HeLa cells as well, microscopically detectable P-bodies are not required for NMD (paper VI). This approach was therefore not suitable to study the function of UPF1 in NMD.

The other approach I started to gain more insight into the function of UPF1 was the biochemical purification of UPF1-bound PTC+ mRNPs. The main technical problem was that I could never achieve high expression levels of the stably expressed UPF1 mutants. In total, I generated in 4 different generations over 16 different HeLa polyclonal cell pools stably expressing the UPF1 mutants, and not one of these pools exhibited high expression levels of UPF1. Possibly, the amount of UPF1 a cell expresses is very tightly regulated. Or the expression of the UPF1 mutants is toxic for the cells, and therefore during the selection with the antibiotic, only cells with low UPF1 expression levels would have survived. However, in transient transfections, it is possible to reach high expression levels of mutant UPF1 already two days after transfection (see Fig 8D). Therefore, this tandem purification strategy with one step pulling at the PTC+ mRNA, and the other step pulling at the mutant UPF1, is only feasible with cell numbers, which can be reached by transient transfections (i.e. up to 10^8 cells). It was therefore necessary to find a way to decrease the required amount of input material for the final mass spectrometry analysis. On the one hand, the sensitivity of the mass spectrometer is a limiting factor, and on the other hand, the necessity to gel-purify the eluates. According to Marc Bühler it should be possible to perform a mass spectrometry analysis without purifying the samples over a gel, given that the eluates have a high purity. To test this, we decided to reduce the complexity of the purification strategy by the use of the RAT-tag. I hope that by the use of the RAT tag, I will be able to purify adequate amounts of mRNPs for the mass spectrometer, to identify and compare the protein-composition of PTC+ and PTC- mRNPs. Once the RAT purification works properly, it should be possible to establish the double-step mRNP purification via MS2MBP and HA-tagged UPF1.

3.3. The physiological role of NMD

Yet, the physiological role of NMD is not understood completely. In yeast, null mutations of Upf1p, Upf2p or Upf3p did not show any phenotype, suggesting that NMD is not essential for viability under laboratory conditions (Leeds et al., 1991; Leeds et al., 1992; Cui et al., 1995; He & Jacobson, 1995). In contrast, UPF1 is essential for the development and viability of plants and mammals (Medghalchi et al., 2001; Yoine et al., 2006). However, because UPF1 is also implicated in additional cellular processes such as translation, telomere maintenance and genome stability, it is currently not clear if the lethality is indeed due to NMD inactivation (Maderazao et al., 2000; Wang et al., 2001; Nott et al. 2004; Azzalin and Lingner, 2006; Azzalin et al., 2007). Furthermore, the coupling of NMD with alternative splicing seems to be widespread among eukaryotes (Lewis et al., 2003). Notably, the overall NMD capacity can vary considerably between different cell lines, tissues, and individuals (Resta et al., 2006; Linde et al., 2007), and recently, variability in NMD efficiency between cell lines has been reported to correlate with intracellular RNPS1 concentrations (Viegas et al., 2007). Genome-wide transcriptome profiling in yeast, drosophila and human cells revealed that 3-10 % of all mRNAs are regulated by NMD (He et al., 2003; Mendell et al., 2004; Rehwinkel et al., 2005; Wittmann et al., 2006; Weischenfeldt et al., 2008). The population of those transcripts is not only restricted to faulty mRNAs, but also comprises numerous endogenous, physiological transcripts (see paper V table 1). In summary, this suggests that NMD has evolved not only as surveillance process for faulty mRNAs, but as general regulator of post-transcriptional gene regulation.

In the light of this and based on our unified model for NMD, we proposed a novel mode of translation-dependent posttranscriptional gene regulation that involves NMD (papers III and IV). Many mammalian 3' UTRs comprise thousands of nucleotides, are most probably highly structured, and provide binding sites for RNA binding proteins that are known to regulate translation and mRNA stability. Our unified NMD model predicts that in the absence of termination promoting signals, an mRNA with a long 3'UTR will be targeted by NMD. But, such endogenous mRNAs with long 3'UTRs are not necessarily an NMD substrate, because the 3'UTR might include binding sites for factors that either intrinsically have stabilizing effects or can reduce the physical distance between the poly(A) tail and the termination codon to stabilize the mRNA. RNA-binding factors can be either proteins or RNAs, and they may alter the 3D configuration of the 3'UTR by masking mRNA sequences otherwise engaged in intramolecular base pairing, or by interacting with each other and thereby looping out mRNA sequences in-between. Importantly, such protein–protein and protein–RNA interactions can be regulated by environmental cues through signal transduction pathways that modify the involved RNA-binding proteins. We would predict that the proposed mechanism depends on translation and UPF1, and that the population of transcripts affected by Upf1 depletion varies in a tissue-specific manner, during development and differentiation, and by environmental cues in general. Therefore, this mechanism could explain the large number of transcripts regulated by NMD, many of which would not be predicted to be NMD substrates according to the currently prevailing model. And, this mechanism provides a possible explanation for the poor overlap between the sets of NMD-regulated transcripts in different microarray studies on UPF1-depleted or UPF2-depleted cells (Mendell et al., 2004; Wittmann et al., 2006; Weischenfeldt et al., 2008), for the varying efficiency of NMD on different substrates, and for the regulatory role of NMD on physiological mRNAs.

To our knowledge, no physiological examples for this postulated mechanism have been reported until today. A recent study indicates such a translation- and NMD-dependent control of mRNA stability regulating spatial and temporal restricted protein synthesis in synapses of neurons (Giorgi et al., 2007). mRNAs that are alternatively polyadenylated might be an other source for transcripts regulated by this mechanism. The cyclooxygenase-2 (COX-2; NM_000963) gene for instance, is expressed at various levels in different tissues and displays a 2.8-kb and a 4.6-kb mRNA isoform due to the usage of an alternative polyadenylation site. Notably, the 4.6-kb COX-2 mRNA is an unstable short-lived mRNA with a half-life of no more than 3.5 h (Lukiw & Bazan, 1997). The long variant of the COX-2 3'UTR contains several AU-rich elements (AREs), and several regions contributing to mRNA stability and translation efficiency have been mapped (Cok & Morrison, 2001). Not all elements contributing to mRNA stability contain an ARE, and therefore NMD may also participate in the complex posttranscriptional regulation of the COX-2 mRNA. Another interesting candidate is the CD44 mRNA (NM_000610). HeLa cells were found to express three CD44 transcripts of 1.6, 2.0 and 5.0 kb in length due to the use of alternative polyadenylation signals (Vikesaa et al., 2006). The 5 kb CD44 mRNA contains several binding sites for IMP proteins (IMP1, IMP2 and IMP3; IMP1 is a synonym of "insulin-like growth factor 2 mRNA-binding protein 1"), and RNAi-mediated depletion of IMP1 or IMP3 results in destabilization of this mRNA species (Vikesaa et al., 2006). Furthermore, the loss of IMPs leads to the destabilization of mRNAs encoding extracellular matrix and adhesion proteins, and these transcripts are present in IMP RNP granules, suggesting that IMPs are directly involved in the post-transcriptional control of these transcripts. IMP1 can bind to AREs, interacts with PABPC1 (Patel & Bag, 2006), and IMPs were found to form homo- and heterodimers and to bind RNA in a cooperative manner (Nielsen et al., 2004). In summary, this suggests that CD44 mRNA may be an interesting candidate potentially regulated by our postulated mechanism of post-transcriptional gene regulation, and that IMPs might regulate the stability of CD44 mRNA and of other mRNAs containing IMP-binding sites in the 3'UTR in this manner.

3.4. Therapeutic aspects

The development for treatments of genetic diseases will be one of the biggest challenges of clinical research and modern medicine. Approximately 30 % of genetic disorders are assigned to a PTC, resulting in the suppression of full-length protein synthesis (Holbrook et al., 2004). Gene therapy will greatly facilitate the treatment of genetic diseases caused by a truncation of the ORF, but yet it is still far from achieving clinical success (Parekh-Olmedo et al., 2005; Parekh-Olmedo & Kmiec, 2007). However, currently more promising is the use of drugs to inhibit NMD or to suppress the pathogenic PTC leading to translational readthrough (Kellermayer, 2006). Treatments with caffeine or wortmannin (inhibitors of SMG1) were shown to have beneficial effects on the phenotype of Ullrich's disease (Usuki et al., 2004). Furthermore, the use of the readthrough promoting aminoglycoside gentamicin showed positive effects on cystic fibrosis patients carrying a PTC in the CFTR gene (Wilschanski et al., 2003). Unfortunately, these approaches lack gene specifity and may therefore lead to various side-effects in long-term treatments. PTC124, another readthrough promoting agent, has been developed to reduce the toxicity of aminoglycoside treatments (Welch et al., 2007). One major advance of PTC124 is its PTC-specifity; it only promotes the readthrough at ribosomes stalled at a PTC, but not at normal TCs. PTC124 is currently in phase 2b clinical trials for nonsense-mediated Duchenne and Becker muscular dystrophy (DMD/BMD) and nonsense-mediated cystic fibrosis (www.ptcbio.com and www.clinicaltrials.gov). Depending on the success of those clinical trial studies, PTC124 will be clinically available within few years. Furthermore, the application range of PTC124 may be extended to other PTC-mediated diseases. However, a treatment with PTC124 would not be suitable in the case of mutations affecting the 3'UTR that have the potential to alter the spatial relationship between the TC and the poly(A) tail, such as insertions into the 3'UTR, mutations of the natural stop codon, mutations that destroy poly(A) sites or create cryptic ones, and modification of binding sites for RNA binding proteins. Until today, only few 3'UTR-mediated diseases have been identified, although there may be many, including certain types of cancer. To the best of my knowledge, three diseases comply with the criteria described above: Myotonic dystrophy (DM; MIM 160900), α-thalassemia (MIM 141850) and Fukuyama-type congenital muscular dystrophy (FCMD; MIM 253800).

DM is caused by an expanded number of trinucleotide (CTG) repeats in the 3'UTR of a cAMP-dependent protein kinase gene (DMPK; MIM 605377.0001). However, the mutated DMPK transcript is not an NMD substrate (Reto Michel, unpublished observation), but the DMPK transcripts are retained within the nucleus, leading to impaired kinase synthesis (Davis et al., 1997).

The most frequent allele causing α-thalassemia carries an anti-termination mutation of UAA to CAA ($α^{CS}$-globin; MIM 141850.0001) that allows translating ribosomes to proceed into the 3'UTR, which leads to a reduced half-live of the mRNA. Different C-rich regions have been mapped in the α-globin 3'UTR that interact with a ribonucleoprotein complex to mediate the stability of the α-globin mRNA (Wang et al., 1995). It is thought that the loss of this interaction leads to an accelerated shortening of the poly(A) tail and finally to the premature degradation of the aberrant α-globin mRNA by REMD (see paragraph 2.2.; (Morales et al., 1997; Kong & Liebhaber, 2007)). Importantly, the stability of the $α^{CS}$-globin mRNA is UPF1-independent (Kong & Liebhaber, 2007).

87 % of the FCMD-patients carry a 3 kb retrotransposon insertion in the 3'UTR of the fukutin gene ((Kobayashi et al., 1998); MIM 607440.0001), which leads to low FCMD fukutin mRNA levels due to yet unknown reasons. HeLa cells express a fukutin mRNA variant with already a 5 kb long 3'UTR, and interestingly this mRNA is not an NMD substrate (Fig. 11), suggesting that this 3'UTR contains a stabilizing element. A possible explanation for the low fukutin FCMD mRNA levels would be given, if the stabilizing element locates 3' of the retrotransposon insertion site, and therefore the usage of the poly(A) site within the retrotransposon would truncate the stabilizing element. This would in turn result in an mRNA where the TC and the poly(A) tail potentially are distal, and according to our unified model for NMD this mRNA would be an NMD substrate. Therefore, the FCMD mRNA was suggested to be the perfect target for testing our oligonucleotide based approach on a disease-relevant mRNA. Unfortunately, the fukutin minigene I cloned did not behave like the endogenous HeLa fukutin mRNA, as it was expressed at very low levels and responded strongly to a knock-down of UPF1. Therefore, I could not test this

hypothesis yet, but fukutin will be subject to further investigations in the future.

3.5. NMTGS – still enigmatic

The transcriptional state of several endogenous NMD substrates, including Ig-μ (Jack et al., 1989; Muhlemann et al., 2001), had already been assessed, and in all these cases transcription was not affected by the presence of a PTC. It was therefore intriguing and unexpected to find that transcriptional silencing was involved in the strong downregulation of our Ig-μ and Ig-γ minigenes in HeLa cells (Buhler et al., 2005). Strikinlgy, NMTGS is PTC-specific and consistently, I could show that NMTGS depends on the translation of its cognate mRNA (paper II). The apparent coupling of a translational event (the recognition of a PTC) with transcription immediately raises several mechanistic questions: What is the trigger that signals from translation back to transcription, and can this trigger act in *trans*? The overexpression of the siRNAse 3'hExo could suppress NMTGS, suggesting that siRNAs could be involved. However, attempts to detect Ig-μ or Ig-γ specific siRNAs were unsuccessful so far (O.M. and R. Gudipati, unpublished observations). Furthermore, I did not find evidence that NMTGS can be induced in *trans* (Fig. 14). Certainly, these results are preliminary and several controls were lacking. Nevertheless, a continuation of these experiments confirmed that transcriptional silencing of an Ig-μWT reporter gene cannot be induced in *trans* by the expression of a Ig-μter310 reporter gene (R. Joncourt, unpublished observation). In contrast, PTC-specific effects at the site of transcription of the Ig-μ gene have already been observed, suggesting the existence of *cis*-acting elements. First, endogenously expressed PTC-containing Ig-μ transcripts were observed to accumulate near or at the site of transcription (Muhlemann et al., 2001). Second, by using fluorescence recovery after photobleaching (FRAP) techniques, kinetic differences at the site of transcription have been observed between a cell line stably expressing the Ig-μWT minigene and a cell line stably expressing the Ig-μter310 minigene (V. de Turris and Oliver Mühlemann, unpublished observation). This would lead to a model, where the slow transcription kinetics of the PTC+ construct leads to retention of the transcript at the transcription site, which may provide the opportunity for chromatin modifying enzymes to silence the gene. Consistently, I did not find significant differences in the occupancy of polymerase II in cells expressing the Ig-μter310 IRE reporter gene and in cells expressing the Ig-μter310 mutIRE reporter gene (Fig. 13). Furthermore, the transcription kinetics of both PTC+ and PTC- Ig-μ constructs is UPF1-dependent, as the transcription kinetics of the PTC+ and the PTC- Ig-μ constructs changed upon RNAi-mediated depletion of UPF1. Importantly, the transcription kinetics of the PTC- and the PTC+ construct was similar after the knock-down of UPF1 (V. de Turris and Oliver Mühlemann, unpublished observation) and it will be interesting to see, if the difference in transcription kinetics between PTC+ and PTC- constructs is also dependent on the translation of the cognate mRNA.

Another interesting point is the coupling of NMD and NMTGS. The findings that both mechanisms are PTC specific and depend on UPF1 suggest that they rely on the same machinery for PTC recognition and that NMTGS branches from the NMD pathway after UPF1 function. Additionally, the fact that the transcriptional silencend state of the Ig-μter310 mutIRE reporter gene can be reversed by RNAi-mediated depletion of UPF1 (paper II) suggests that UPF1 is not only required to establish NMTGS but also for maintaining it. Moreover, it would be interesting to test, which of the other NMD factors are required for NMTGS too. An especially interesting candidate would be SMG6, because the involvement of SMG6 in NMTGS would suggest that the degradation of the cognate mRNA is required for NMTGS.

Yet, a physiological relevance of NMTGS has not been found. So far, NMTGS was only observed in poly-clonal cell pools. The poly-clonal cell pools we have used represent a very ill defined system, because the constructs are randomly integrated and therefore number of integrated genes and the integration sites vary between the cells. It will be crucial for the further investigation of NMTGS to generate mono-clonal cell lines with the reporter gene integrated once at a specified place. Furthermore, NMTGS was only observed with PTC-containing Ig-µ and Ig-γ minigenes, but not with a Ig-κ minigene or other NMD reporter genes (Buhler et al., 2005), which led to the appealing hypothesis that NMTGS may be required for the silencing of non-productively rearranged heavy-chain alleles. However, no differences in the transcription of productively and non-productively rearranged heavy chain alleles was observed in clonal lines of immortalized murine pro-B cells (A. B. Eberle and O. Mühlemann, manuscript in preparation). Thus, until NMTGS is not observed in a more physiological system or on an endogenous gene, it can not be excluded that NMTGS simply represents an artifact observed only in our poly-clonal cell pools.

4. Materials and Methods

The used materials and methods are essentially described in the attached papers. In this section I will only describe the cloning of plasmids that are not described elsewhere, and the mRNP purification procedures.

4.1. Plasmids

4.1.1. Cloning of pSupuro GW182

pSUPERpuro-GW182 was generated by insertion of double-stranded oligos encoding for shRNAs into pSUPERpuro between the BglII and HindIII sites as described previously (Brummelkamp et al., 2002). The target sequence of GW182 was 5'-GAAATGCTCT-GGTCCGCTA-3' (Lian et al., 2007).

4.1.2. Cloning of the fukutin minigene

Fukutin exon 10 was PCR amplified using the Long-Range PCR kit (Roche) with primers flanked with XhoI and NheI sites (forward: 5'-ACACTCGAGATACCTGTT-TCCGAAGTTTACAC-3' and reverse: 5'-ACAGCTAGCCATCATCCCAAACTGGAT-TATATATTAC-3') out of SalI digested genomic DNA from HeLa cells. This 6 kb long fragment was then ligated into a XbaI and XhoI digested and dephosphorylated pcDNA3-HA vector (Invitrogen), resulting in pcDNA3-HA-fukutin_ex10WT. A fragment containing fukutin exons 8 and 9 and the full-length intron was PCR amplified using the PCR FastTaq Mastermix (Roche) with primers flanked with XhoI sites (forward: 5'-ACAGTCGACATCGATGGATGGTATCGACAATGCAACATTATTCC-3' and reverse: 5'- ACACTCGAGTTGAATTTTTTTCCTGTTTTGGCCTGAG-3') out of SalI digested genomic DNA from HeLa cells. This 2 kb long fragment was ligated into XhoI digested pcDNA3-HA-fukutin_ex10WT, resulting in the vector pcDNA3-HA-fukutinWT. All sequences were verified by sequencing.

4.2. mRNP purifications

4.2.1. Double step purification with MS2MBP and α-HA antibodies

All purification steps were performed at 4 °C. 10^7 HeLa cells were lysed in 500 µl column buffer (CB; 20 mM TrisHCl pH 7.4, 200 mM NaCl, 1 mM EDTA, in DEPC-treated H_2O) containing 0.5 % Triton X-100 and sonicated 3 times for 10 s. The lysate was centrifuged 10 min at 13'000 rpm, the supernatant was added to 100 µl pre-washed amylose beads (New England Biolabs) in a final volume of 1.5 ml CB and rotated for 5 h. The beads were washed 5 times with CB containing 150 mM NaCl and eluted 3 times for 10 min in a total volume of 600 µl with CB containing 50 mM maltose. 500 µl of the eluate was added to 100 µl protein G-sepharose beads (GE Healthcare), which have been pre-incubated with 10 µl rabbit α-HA antibodies (Y11, Santa Cruz) for 1 h and washed twice in HA equilibration buffer (HA-Eq; 20 mM Tris pH 7.5; 150 mM NaCl, 0.1 mM EDTA) in a final volume of 1.5 ml HA-Eq. The beads were rotated over-night, washed 6 times with HA-Eq-buffer containing 0.05 % Tween20 and eluted 3 times for 15 minutes shaking at 600 rpm at 37 °C in a total volume of 500 µl elution buffer (1 mg/ml HA-peptide, 2 % RNasIn (Promega), Protease Inhibitors EDTA-free (Roche), in HA-Eq-buffer).

4.2.2. RAT purification

All purification steps were performed at 4 °C. The cells were washed in PBS and lysed in PP7-BP (20 mM Hepes pH 7.9, 150 mM NaCl, 2 mM $MgCl_2$, 10 % glycerol, 1 mM DTT) containing EDTA-free protease inhibitors (Roche) and 0.5 % NP-40, sonicated 3 times for 10 s and centrifuged 10 min at 13'000 rpm. The lysate was diluted to 2.5 mg/ml protein and 3.5 µg/ml PP7 coat protein was added together with the radioactive labeled βGl probe. The lysate was rotated for 60 min, then 100 µl/ml rabbit-IgG agarose beads (Sigma) were added, and rotated for 90 min. The beads were washed twice with PP7-BP containing EDTA-free protease inhibitors and 0.1 % NP-40 and once with PP7-BP, and eluted by 60 min rotation with acTEV protease (20 U, Invitrogen) in a total volume of 100 µl. The rabbit-IgG agarose beads were washed once with 100 µl PP7-BP, and the wash fraction was pooled together with the elution, and 30 µl tobramycin-resin was added, followed by 45 min rotation. Tobramycin-resin was generated by coupling tobramycin (Sigma) to Affi-Gel 10 resin (Bio-Rad). Resin was resuspended in an equal volume of 20 mM Hepes (pH 7.5) and 100 mM tobramycin, rotated overnight at 4°C, and then supplemented with 100 mM ethanolamine (pH 8.0) for 2 h to block unreacted coupling sites. The RNA was eluted twice with 100 µl elution buffer (25 mM tobramycin, 20 mM Hepes pH 7.5) rotating for 10 min at 4 °C.

5. REFERENCES

Alexandrov A, Chernyakov I, Gu W, Hiley SL, Hughes TR, Grayhack EJ, Phizicky EM. 2006. Rapid tRNA decay can result from lack of nonessential modifications. Mol Cell 21:87-96.

Amrani N, Ganesan R, Kervestin S, Mangus DA, Ghosh S, Jacobson A. 2004. A faux 3'-UTR promotes aberrant termination and triggers nonsense-mediated mRNA decay. Nature 432:112-118.

Amrani N, Sachs MS, Jacobson A. 2006. Early nonsense: mRNA decay solves a translational problem. Nat Rev Mol Cell Biol 7:415-425.

Anders KR, Grimson A, Anderson P. 2003. SMG-5, required for C.elegans nonsense-mediated mRNA decay, associates with SMG-2 and protein phosphatase 2A. EMBO J 22:641-650.

Arciga-Reyes L, Wootton L, Kieffer M, Davies B. 2006. UPF1 is required for nonsense-mediated mRNA decay (NMD) and RNAi in Arabidopsis. The Plant Journal 47:480-489.

Azzalin CM, Lingner J. 2006. The human RNA surveillance factor UPF1 is required for S phase progression and genome stability. Curr Biol 16:433-439.

Azzalin CM, Reichenbach P, Khoriauli L, Giulotto E, Lingner J. 2007. Telomeric repeat containing RNA and RNA surveillance factors at mammalian chromosome ends. Science 318:798-801.

Behm-Ansmant I, Gatfield D, Rehwinkel J, Hilgers V, Izaurralde E. 2007. A conserved role for cytoplasmic poly(A)-binding protein 1 (PABPC1) in nonsense-mediated mRNA decay. EMBO J.

Behm-Ansmant I, Kashima I, Rehwinkel J, Sauliere J, Wittkopp N, Izaurralde E. 2007. mRNA quality control: An ancient machinery recognizes and degrades mRNAs with nonsense codons. FEBS Lett 581:2845-2853.

Bhattacharya A, Czaplinski K, Trifillis P, He F, Jacobson A, Peltz SW. 2000. Characterization of the biochemical properties of the human Upf1 gene product that is involved in nonsense-mediated mRNA decay. RNA 6:1226-1235.

Bloch DB, Gulick T, Bloch KD, Yang WH. 2006. Processing body autoantibodies reconsidered. RNA 12:707-709.

Brumbaugh KM, Otterness DM, Geisen C, Oliveira V, Brognard J, Li X, Lejeune F,

References

Tibbetts RS, Maquat LE, Abraham RT. 2004. The mRNA surveillance protein hSMG-1 functions in genotoxic stress response pathways in mammalian cells. Mol Cell 14:585-598.

Brummelkamp TR, Bernards R, Agami R. 2002. A system for stable expression of short interfering RNAs in mammalian cells. Science 296:550-553.

Buhler M, Mohn F, Stalder L, Muhlemann O. 2005. Transcriptional silencing of nonsense codon-containing immunoglobulin minigenes. Mol Cell 18:307-317.

Buhler M, Paillusson A, Muhlemann O. 2004. Efficient downregulation of immunoglobulin mu mRNA with premature translation-termination codons requires the 5'-half of the VDJ exon. Nucleic Acids Res 32:3304-3315.

Buhler M, Steiner S, Mohn F, Paillusson A, Muhlemann O. 2006. EJC-independent degradation of nonsense immunoglobulin-mu mRNA depends on 3' UTR length. Nat Struct Mol Biol 13:462-464.

Buhler M, Wilkinson MF, Muhlemann O. 2002. Intranuclear degradation of nonsense codon-containing mRNA. EMBO Rep 3:646-651.

Buzina A, Shulman MJ. 1999. Infrequent translation of a nonsense codon is sufficient to decrease mRNA level. Mol Biol Cell 10:515-524.

Cao D, Parker R. 2003. Computational modeling and experimental analysis of nonsense-mediated decay in yeast. Cell 113:533-545.

Carninci P, Kasukawa T, Katayama S, Gough J, Frith MC, Maeda N, Oyama R, Ravasi T, Lenhard B, Wells C, Kodzius R, Shimokawa K, Bajic VB, Brenner SE, Batalov S, Forrest AR, Zavolan M, Davis MJ, Wilming LG, Aidinis V, Allen JE, Ambesi-Impiombato A, Apweiler R, Aturaliya RN, Bailey TL, Bansal M, Baxter L, Beisel KW, Bersano T, Bono H, Chalk AM, Chiu KP, Choudhary V, Christoffels A, Clutterbuck DR, Crowe ML, Dalla E, Dalrymple BP, de Bono B, Della Gatta G, di Bernardo D, Down T, Engstrom P, Fagiolini M, Faulkner G, Fletcher CF, Fukushima T, Furuno M, Futaki S, Gariboldi M, Georgii-Hemming P, Gingeras TR, Gojobori T, Green RE, Gustincich S, Harbers M, Hayashi Y, Hensch TK, Hirokawa N, Hill D, Huminiecki L, Iacono M, Ikeo K, Iwama A, Ishikawa T, Jakt M, Kanapin A, Katoh M, Kawasawa Y, Kelso J, Kitamura H, Kitano H, Kollias G, Krishnan SP, Kruger A, Kummerfeld SK, Kurochkin IV, Lareau LF, Lazarevic D, Lipovich L, Liu J, Liuni S, McWilliam S, Madan Babu M, Madera M, Marchionni L, Matsuda H, Matsuzawa S, Miki H, Mignone F, Miyake S, Morris K, Mottagui-Tabar S, Mulder N, Nakano N, Nakauchi H, Ng P, Nilsson R, Nishiguchi S, Nishikawa S, Nori F, Ohara O, Okazaki Y, Orlando V, Pang KC, Pavan WJ, Pavesi G, Pesole G, Petrovsky N, Piazza S, Reed J, Reid JF, Ring BZ, Ringwald M, Rost B, Ruan Y, Salzberg SL, Sandelin A, Schneider C, Schonbach C, Sekiguchi K, Semple CA, Seno S, Sessa L, Sheng Y, Shibata Y, Shimada H, Shimada K, Silva D, Sinclair B, Sperling

S, Stupka E, Sugiura K, Sultana R, Takenaka Y, Taki K, Tammoja K, Tan SL, Tang S, Taylor MS, Tegner J, Teichmann SA, Ueda HR, van Nimwegen E, Verardo R, Wei CL, Yagi K, Yamanishi H, Zabarovsky E, Zhu S, Zimmer A, Hide W, Bult C, Grimmond SM, Teasdale RD, Liu ET, Brusic V, Quackenbush J, Wahlestedt C, Mattick JS, Hume DA, Kai C, Sasaki D, Tomaru Y, Fukuda S, Kanamori-Katayama M, Suzuki M, Aoki J, Arakawa T, Iida J, Imamura K, Itoh M, Kato T, Kawaji H, Kawagashira N, Kawashima T, Kojima M, Kondo S, Konno H, Nakano K, Ninomiya N, Nishio T, Okada M, Plessy C, Shibata K, Shiraki T, Suzuki S, Tagami M, Waki K, Watahiki A, Okamura-Oho Y, Suzuki H, Kawai J, Hayashizaki Y. 2005. The transcriptional landscape of the mammalian genome. Science 309:1559-1563.

Carter MS, Li S, Wilkinson MF. 1996. A splicing-dependent regulatory mechanism that detects translation signals. Embo J 15:5965-5975.

Cavalier-Smith T. 1991. Intron phylogeny: a new hypothesis. Trends Genet 7:145-148.

Chan WK, Huang L, Gudikote JP, Chang YF, Imam JS, MacLean JA, 2nd, Wilkinson MF. 2007. An alternative branch of the nonsense-mediated decay pathway. EMBO J 26:1820-1830.

Chang JC, Kan YW. 1979. beta 0 thalassemia, a nonsense mutation in man. Proc Natl Acad Sci U S A 76:2886-2889.

Chang YF, Imam JS, Wilkinson MF. 2007. The nonsense-mediated decay RNA surveillance pathway. Annu Rev Biochem 76:51-74.

Chen CY, Shyu AB. 2003. Rapid deadenylation triggered by a nonsense codon precedes decay of the RNA body in a mammalian cytoplasmic nonsense-mediated decay pathway. Mol Cell Biol 23:4805-4813.

Cheng Z, Muhlrad D, Lim MK, Parker R, Song H. 2007. Structural and functional insights into the human Upf1 helicase core. EMBO J 26:253-264.

Chiu S-Y, Serin G, Ohara O, Maquat LE. 2003. Characterization of human Smg5/7a: A protein with similarities to Caenorhabditis elegans SMG5 and SMG7 that functions in the dephosphorylation of Upf1. RNA 9:77-87.

Cok SJ, Morrison AR. 2001. The 3'-untranslated region of murine cyclooxygenase-2 contains multiple regulatory elements that alter message stability and translational efficiency. J Biol Chem 276:23179-23185.

Connor A, Wiersma E, Shulman MJ. 1994. On the linkage between RNA processing and RNA translatability. J Biol Chem 269:25178-25184.

Cougot N, Babajko S, Seraphin B. 2004. Cytoplasmic foci are sites of mRNA decay in human cells. J Cell Biol 165:31-40.

Cui Y, Hagan KW, Zhang S, Peltz SW. 1995. Identification and characterization of genes that are required for the accelerated degradation of mRNAs containing a premature translational termination codon. Genes Dev 9:423-436.

Czaplinski K, Ruiz-Echevarria MJ, Paushkin SV, Han X, Weng Y, Perlick HA, Dietz HC, Ter-Avanesyan MD, Peltz SW. 1998. The surveillance complex interacts with the translation release factors to enhance termination and degrade aberrant mRNAs. Genes Dev 12:1665-1677.

Davis BM, McCurrach ME, Taneja KL, Singer RH, Housman DE. 1997. Expansion of a CUG trinucleotide repeat in the 3' untranslated region of myotonic dystrophy protein kinase transcripts results in nuclear retention of transcripts. Proc Natl Acad Sci U S A 94:7388-7393.

Doma MK, Parker R. 2006. Endonucleolytic cleavage of eukaryotic mRNAs with stalls in translation elongation. Nature 440:561-564.

Doma MK, Parker R. 2007. RNA quality control in eukaryotes. Cell 131:660-668.

Domeier ME, Morse DP, Knight SW, Portereiko M, Bass BL, Mango SE. 2000. A link between RNA interference and nonsense-mediated decay in Caenorhabditis elegans. Science 289:1928-1931.

Durand S, Cougot N, Mahuteau-Betzer F, Nguyen CH, Grierson DS, Bertrand E, Tazi J, Lejeune F. 2007. Inhibition of nonsense-mediated mRNA decay (NMD) by a new chemical molecule reveals the dynamic of NMD factors in P-bodies. J Cell Biol 178:1145-1160.

Eberle BA, Stalder L, Mathys H, Zamudio R, Muhlemann O. 2008. Posttranscriptional gene regulation by spatial rearrangement of the 3' untranslated region. PLoS Biol 6:e92.

Eulalio A, Behm-Ansmant I, Izaurralde E. 2007. P bodies: at the crossroads of posttranscriptional pathways. Nat Rev Mol Cell Biol 8:9.

Eulalio A, Behm-Ansmant I, Schweizer D, Izaurralde E. 2007. P-Body Formation Is a Consequence, Not the Cause, of RNA-Mediated Gene Silencing. Mol Cell Biol 27:3970-3981.

Fenger-Gron M, Fillman C, Norrild B, Lykke-Andersen J. 2005. Multiple Processing Body Factors and the ARE Binding Protein TTP Activate mRNA Decapping. Mol Cell 20:905.

Frischmeyer PA, van Hoof A, O'Donnell K, Guerrerio AL, Parker R, Dietz HC. 2002. An mRNA surveillance mechanism that eliminates transcripts lacking termination codons. Science 295:2258-2261.

Frith MC, Pheasant M, Mattick JS. 2005. The amazing complexity of the human transcriptome. Eur J Hum Genet 13:894-897.

Fukuhara N, Ebert J, Unterholzner L, Lindner D, Izaurralde E, Conti E. 2005. SMG7 is a 14-3-3-like adaptor in the nonsense-mediated mRNA decay pathway. Mol Cell 17:537-547.

Gagen MJ, Mattick JS. 2005. Inherent size constraints on prokaryote gene networks due to „accelerating" growth. Theory Biosci 123:381-411.

Gatfield D, Izaurralde E. 2004. Nonsense-mediated messenger RNA decay is initiated by endonucleolytic cleavage in Drosophila. Nature 429:575-578.

Gatfield D, Unterholzner L, Ciccarelli FD, Bork P, Izaurralde E. 2003. Nonsense-mediated mRNA decay in Drosophila: at the intersection of the yeast and mammalian pathways. EMBO J 22:3960-3970.

Gehring NH, Kunz JB, Neu-Yilik G, Breit S, Viegas MH, Hentze MW, Kulozik AE. 2005. Exon-junction complex components specify distinct routes of nonsense-mediated mRNA decay with differential cofactor requirements. Mol Cell 20:65-75.

Gehring NH, Neu-Yilik G, Schell T, Hentze MW, Kulozik AE. 2003. Y14 and hUpf3b form an NMD-activating complex. Mol Cell 11:939-949.

Giorgi C, Yeo GW, Stone ME, Katz DB, Burge C, Turrigiano G, Moore MJ. 2007. The EJC factor eIF4AIII modulates synaptic strength and neuronal protein expression. Cell 130:179-191.

Gonzalez CI, Ruiz-Echevarria MJ, Vasudevan S, Henry MF, Peltz SW. 2000. The yeast hnRNP-like protein Hrp1/Nab4 marks a transcript for nonsense-mediated mRNA decay. Mol Cell 5:489-499.

He F, Jacobson A. 1995. Identification of a novel component of the nonsense-mediated mRNA decay pathway by use of an interacting protein screen. Genes Dev 9:437-454.

He F, Li X, Spatrick P, Casillo R, Dong S, Jacobson A. 2003. Genome-wide analysis of mRNAs regulated by the nonsense-mediated and 5' to 3' mRNA decay pathways in yeast. Mol Cell 12:1439-1452.

Hentze MW, Kuhn LC. 1996. Molecular control of vertebrate iron metabolism: mRNA-based regulatory circuits operated by iron, nitric oxide, and oxidative stress. Proc Natl Acad Sci U S A 93:8175-8182.

Hermann T. 2007. Aminoglycoside antibiotics: old drugs and new therapeutic approaches. Cell Mol Life Sci 64:1841-1852.

Hilleren P, Parker R. 1999. mRNA surveillance in eukaryotes: kinetic proofreading of proper translation termination as assessed by mRNP domain organization? RNA 5:711-719.

Hock J, Weinmann L, Ender C, Rudel S, Kremmer E, Raabe M, Urlaub H, Meister G. 2007. Proteomic and functional analysis of Argonaute-containing mRNA-protein complexes in human cells. EMBO Rep 8:1052-1060.

Hogg JR, Collins K. 2007. RNA-based affinity purification reveals 7SK RNPs with distinct composition and regulation. Rna 13:868-880.

Holbrook JA, Neu-Yilik G, Hentze MW, Kulozik AE. 2004. Nonsense-mediated decay approaches the clinic. Nat Genet 36:801-808.

Iborra FJ, Jackson DA, Cook PR. 2001. Coupled transcription and translation within nuclei of mammalian cells. Science 293:1139-1142.

Inada T, Aiba H. 2005. Translation of aberrant mRNAs lacking a termination codon or with a shortened 3'-UTR is repressed after initiation in yeast. Embo J 24:1584-1595.

Ishigaki Y, Li X, Serin G, Maquat LE. 2001. Evidence for a pioneer round of mRNA translation: mRNAs subject to nonsense-mediated decay in mammalian cells are bound by CBP80 and CBP20. Cell 106:607-617.

Isken O, Maquat LE. 2007. Quality control of eukaryotic mRNA: safeguarding cells from abnormal mRNA function. Genes Dev 21:1833-3856.

Isken O, Maquat LE. 2008. The multiple lives of NMD factors: balancing roles in gene and genome regulation. Nat Rev Genet.

Ivanov PV, Gehring NH, Kunz JB, Hentze MW, Kulozik AE. 2008. Interactions between UPF1, eRFs, PABP and the exon junction complex suggest an integrated model for mammalian NMD pathways. EMBO J.

Jack HM, Berg J, Wabl M. 1989. Translation affects immunoglobulin mRNA stability. Eur J Immunol 19:843-847.

Kadlec J, Guilligay D, Ravelli RB, Cusack S. 2006. Crystal structure of the UPF2-interacting domain of nonsense-mediated mRNA decay factor UPF1. RNA 12:1817-1824.

Kashima I, Yamashita A, Izumi N, Kataoka N, Morishita R, Hoshino S, Ohno M, Dreyfuss G, Ohno S. 2006. Binding of a novel SMG-1-Upf1-eRF1-eRF3 complex (SURF) to the exon junction complex triggers Upf1 phosphorylation and nonsense-mediated mRNA decay. Genes Dev 20:355-367.

Kaygun H, Marzluff WF. 2005. Regulated degradation of replication-dependent histone mRNAs requires both ATR and Upf1. Nat Struct Mol Biol 12:794-800.

Kellermayer R. 2006. Translational readthrough induction of pathogenic nonsense mutations. Eur J Med Genet 49:445-450.

Kerr TP, Sewry CA, Robb SA, Roberts RG. 2001. Long mutant dystrophins and variable phenotypes: evasion of nonsense-mediated decay? Hum Genet 109:402-407.

Kertesz S, Kerenyi Z, Merai Z, Bartos I, Palfy T, Barta E, Silhavy D. 2006. Both introns and long 3'-UTRs operate as cis-acting elements to trigger nonsense-mediated decay in plants. Nucleic Acids Res 34:6147-6157.

Kim JK, Gabel HW, Kamath RS, Tewari M, Pasquinelli A, Rual J-F, Kennedy S, Dybbs M, Bertin N, Kaplan JM, Vidal M, Ruvkun G. 2005. Functional Genomic Analysis of RNA Interference in C. elegans. Science 308:1164-1167.

Kim YK, Furic L, Desgroseillers L, Maquat LE. 2005. Mammalian Staufen1 recruits Upf1 to specific mRNA 3'UTRs so as to elicit mRNA decay. Cell 120:195-208.

Klausner RD, Rouault TA, Harford JB. 1993. Regulating the fate of mRNA: the control of cellular iron metabolism. Cell 72:19-28.

Kobayashi K, Nakahori Y, Miyake M, Matsumura K, Kondo-Iida E, Nomura Y, Segawa M, Yoshioka M, Saito K, Osawa M, Hamano K, Sakakihara Y, Nonaka I, Nakagome Y, Kanazawa I, Nakamura Y, Tokunaga K, Toda T. 1998. An ancient retrotransposal insertion causes Fukuyama-type congenital muscular dystrophy. Nature 394:388.

Kobayashi K, Sasaki J, Kondo-Iida E, Fukuda Y, Kinoshita M, Sunada Y, Nakamura Y, Toda T. 2001. Structural organization, complete genomic sequences and mutational analyses of the Fukuyama-type congenital muscular dystrophy gene, fukutin. FEBS Lett 489:192-196.

Kong J, Liebhaber SA. 2007. A cell type-restricted mRNA surveillance pathway triggered by ribosome extension into the 3' untranslated region. Nat Struct Mol Biol 14:670-676.

References

Kugler W, Enssle J, Hentze MW, Kulozik AE. 1995. Nuclear degradation of nonsense mutated beta-globin mRNA: a post-transcriptional mechanism to protect heterozygotes from severe clinical manifestations of beta-thalassemia? Nucleic Acids Res 23:413-418.

Lander ES, Linton LM, Birren B, Nusbaum C, Zody MC, Baldwin J, Devon K, Dewar K, Doyle M, FitzHugh W, Funke R, Gage D, Harris K, Heaford A, Howland J, Kann L, Lehoczky J, LeVine R, McEwan P, McKernan K, Meldrim J, Mesirov JP, Miranda C, Morris W, Naylor J, Raymond C, Rosetti M, Santos R, Sheridan A, Sougnez C, Stange-Thomann N, Stojanovic N, Subramanian A, Wyman D, Rogers J, Sulston J, Ainscough R, Beck S, Bentley D, Burton J, Clee C, Carter N, Coulson A, Deadman R, Deloukas P, Dunham A, Dunham I, Durbin R, French L, Grafham D, Gregory S, Hubbard T, Humphray S, Hunt A, Jones M, Lloyd C, McMurray A, Matthews L, Mercer S, Milne S, Mullikin JC, Mungall A, Plumb R, Ross M, Shownkeen R, Sims S, Waterston RH, Wilson RK, Hillier LW, McPherson JD, Marra MA, Mardis ER, Fulton LA, Chinwalla AT, Pepin KH, Gish WR, Chissoe SL, Wendl MC, Delehaunty KD, Miner TL, Delehaunty A, Kramer JB, Cook LL, Fulton RS, Johnson DL, Minx PJ, Clifton SW, Hawkins T, Branscomb E, Predki P, Richardson P, Wenning S, Slezak T, Doggett N, Cheng JF, Olsen A, Lucas S, Elkin C, Uberbacher E, Frazier M, Gibbs RA, Muzny DM, Scherer SE, Bouck JB, Sodergren EJ, Worley KC, Rives CM, Gorrell JH, Metzker ML, Naylor SL, Kucherlapati RS, Nelson DL, Weinstock GM, Sakaki Y, Fujiyama A, Hattori M, Yada T, Toyoda A, Itoh T, Kawagoe C, Watanabe H, Totoki Y, Taylor T, Weissenbach J, Heilig R, Saurin W, Artiguenave F, Brottier P, Bruls T, Pelletier E, Robert C, Wincker P, Smith DR, Doucette-Stamm L, Rubenfield M, Weinstock K, Lee HM, Dubois J, Rosenthal A, Platzer M, Nyakatura G, Taudien S, Rump A, Yang H, Yu J, Wang J, Huang G, Gu J, Hood L, Rowen L, Madan A, Qin S, Davis RW, Federspiel NA, Abola AP, Proctor MJ, Myers RM, Schmutz J, Dickson M, Grimwood J, Cox DR, Olson MV, Kaul R, Raymond C, Shimizu N, Kawasaki K, Minoshima S, Evans GA, Athanasiou M, Schultz R, Roe BA, Chen F, Pan H, Ramser J, Lehrach H, Reinhardt R, McCombie WR, de la Bastide M, Dedhia N, Blocker H, Hornischer K, Nordsiek G, Agarwala R, Aravind L, Bailey JA, Bateman A, Batzoglou S, Birney E, Bork P, Brown DG, Burge CB, Cerutti L, Chen HC, Church D, Clamp M, Copley RR, Doerks T, Eddy SR, Eichler EE, Furey TS, Galagan J, Gilbert JG, Harmon C, Hayashizaki Y, Haussler D, Hermjakob H, Hokamp K, Jang W, Johnson LS, Jones TA, Kasif S, Kaspryzk A, Kennedy S, Kent WJ, Kitts P, Koonin EV, Korf I, Kulp D, Lancet D, Lowe TM, McLysaght A, Mikkelsen T, Moran JV, Mulder N, Pollara VJ, Ponting CP, Schuler G, Schultz J, Slater G, Smit AF, Stupka E, Szustakowski J, Thierry-Mieg D, Thierry-Mieg J, Wagner L, Wallis J, Wheeler R, Williams A, Wolf YI, Wolfe KH, Yang SP, Yeh RF, Collins F, Guyer MS, Peterson J, Felsenfeld A, Wetterstrand KA, Patrinos A, Morgan MJ, de Jong P, Catanese JJ, Osoegawa K, Shizuya H, Choi S, Chen YJ. 2001. Initial sequencing and analysis of the human genome. Nature 409:860-921.

LaRiviere FJ, Cole SE, Ferullo DJ, Moore MJ. 2006. A late-acting quality control process for mature eukaryotic rRNAs. Mol Cell 24:619-626.

Le Hir H, Andersen GR. 2008. Structural insights into the exon junction complex. Curr Opin Struct Biol 18:112-119.

Leeds P, Peltz SW, Jacobson A, Culbertson MR. 1991. The product of the yeast UPF1 gene is required for rapid turnover of mRNAs containing a premature translational termination codon. Genes Dev 5:2303-2314.

Leeds P, Wood JM, Lee BS, Culbertson MR. 1992. Gene products that promote mRNA turnover in Saccharomyces cerevisiae. Mol Cell Biol 12:2165-2177.

Lewis BP, Green RE, Brenner SE. 2003. Evidence for the widespread coupling of alternative splicing and nonsense-mediated mRNA decay in humans. Proc Natl Acad Sci U S A 100:189-192.

Lian S, Fritzler MJ, Katz J, Hamazaki T, Terada N, Satoh M, Chan EK. 2007. Small interfering RNA-mediated silencing induces target-dependent assembly of GW/P bodies. Mol Biol Cell 18:3375-3387.

Linde L, Boelz S, Neu-Yilik G, Kulozik AE, Kerem B. 2007. The efficiency of nonsense-mediated mRNA decay is an inherent character and varies among different cells. Eur J Hum Genet 15:1156-1162.

Longman D, Plasterk RH, Johnstone IL, Caceres JF. 2007. Mechanistic insights and identification of two novel factors in the C. elegans NMD pathway. Genes Dev 21:1075-1085.

Lukiw WJ, Bazan NG. 1997. Cyclooxygenase 2 RNA message abundance, stability, and hypervariability in sporadic Alzheimer neocortex. J Neurosci Res 50:937-945.

Lykke-Andersen J. 2002. Identification of a human decapping complex associated with hUpf proteins in nonsense-mediated decay. Mol Cell Biol 22:8114-8121.

Lykke-Andersen J, Shu MD, Steitz JA. 2001. Communication of the position of exon-exon junctions to the mRNA surveillance machinery by the protein RNPS1. Science 293:1836-1839.

Maderazo AB, He F, Mangus DA, Jacobson A. 2000. Upf1p control of nonsense mRNA translation is regulated by Nmd2p and Upf3p. Mol Cell Biol 20:4591-4603.

Maniatis T, Reed R. 2002. An extensive network of coupling among gene expression machines. Nature 416:499-506.

Maquat LE. 2004. Nonsense-mediated mRNA decay: splicing, translation and mRNP dynamics. Nat Rev Mol Cell Biol 5:89-99.

Matsuda D, Hosoda N, Kim YK, Maquat LE. 2007. Failsafe nonsense-mediated mRNA decay does not detectably target eIF4E-bound mRNA. Nat Struct Mol Biol 14:974-979.

References

Mattick JS. 2003. Challenging the dogma: the hidden layer of non-protein-coding RNAs in complex organisms. Bioessays 25:930-939.

Mattick JS. 2004. RNA regulation: a new genetics? Nat Rev Genet 5:316-323.

Mattick JS, Gagen MJ. 2005. MATHEMATICS/COMPUTATION: Accelerating Networks. Science 307:856-858.

Mattick JS, Makunin IV. 2006. Non-coding RNA. Hum Mol Genet 15 Spec No 1:R17-29.

Medghalchi SM, Frischmeyer PA, Mendell JT, Kelly AG, Lawler AM, Dietz HC. 2001. Rent1, a trans-effector of nonsense-mediated mRNA decay, is essential for mammalian embryonic viability. Hum Mol Genet 10:99-105.

Mendell JT, ap Rhys CM, Dietz HC. 2002. Separable roles for rent1/hUpf1 in altered splicing and decay of nonsense transcripts. Science 298:419-422.

Mendell JT, Sharifi NA, Meyers JL, Martinez-Murillo F, Dietz HC. 2004. Nonsense surveillance regulates expression of diverse classes of mammalian transcripts and mutes genomic noise. Nat Genet 36:1073-1078.

Mendes Soares LM, Valcarcel J. 2006. The expanding transcriptome: the genome as the ‚Book of Sand'. Embo J 25:923-931.

Morales J, Russell JE, Liebhaber SA. 1997. Destabilization of human alpha-globin mRNA by translation anti-termination is controlled during erythroid differentiation and is paralleled by phased shortening of the poly(A) tail. J Biol Chem 272:6607-6613.

Muhlemann O, Mock-Casagrande CS, Wang J, Li S, Custodio N, Carmo-Fonseca M, Wilkinson MF, Moore MJ. 2001. Precursor RNAs harboring nonsense codons accumulate near the site of transcription. Mol Cell 8:33-43.

Muhlrad D, Parker R. 1994. Premature translational termination triggers mRNA decapping. Nature 370:578-581.

Muhlrad D, Parker R. 1999. Aberrant mRNAs with extended 3' UTRs are substrates for rapid degradation by mRNA surveillance. RNA 5:1299-1307.

Muller B, Blackburn J, Feijoo C, Zhao X, Smythe C. 2007. DNA-activated protein kinase functions in a newly observed S phase checkpoint that links histone mRNA abundance with DNA replication. J Cell Biol 179:1385-1398.

Nielsen J, Kristensen MA, Willemoes M, Nielsen FC, Christiansen J. 2004. Sequential dimerization of human zipcode-binding protein IMP1 on RNA: a cooperative mechanism providing RNP stability. Nucleic Acids Res 32:4368-4376.

Nott A, Le Hir H, Moore MJ. 2004. Splicing enhances translation in mammalian cells: an additional function of the exon junction complex. Genes Dev 18:210-222.

Ohnishi T, Yamashita A, Kashima I, Schell T, Anders KR, Grimson A, Hachiya T, Hentze MW, Anderson P, Ohno S. 2003. Phosphorylation of hUPF1 induces formation of mRNA surveillance complexes containing hSMG-5 and hSMG-7. Mol Cell 12:1187-1200.

Ou Y, Enarson P, Rattner JB, Barr SG, Fritzler MJ. 2004. The nuclear pore complex protein Tpr is a common autoantigen in sera that demonstrate nuclear envelope staining by indirect immunofluorescence. Clin Exp Immunol 136:379-387.

Page MF, Carr B, Anders KR, Grimson A, Anderson P. 1999. SMG-2 Is a Phosphorylated Protein Required for mRNA Surveillance in Caenorhabditis elegans and Related to Upf1p of Yeast. Mol Cell Biol 19:5943-5951.

Palacios IM, Gatfield D, St Johnston D, Izaurralde E. 2004. An eIF4AIII-containing complex required for mRNA localization and nonsense-mediated mRNA decay. Nature 427:753.

Parekh-Olmedo H, Ferrara L, Brachman E, Kmiec EB. 2005. Gene therapy progress and prospects: targeted gene repair. Gene Ther 12:639-646.

Parekh-Olmedo H, Kmiec EB. 2007. Progress and prospects: targeted gene alteration (TGA). Gene Ther 14:1675-1680.

Parker R, Sheth U. 2007. P bodies and the control of mRNA translation and degradation. Mol Cell 25:635-646.

Patel GP, Bag J. 2006. IMP1 interacts with poly(A)-binding protein (PABP) and the autoregulatory translational control element of PABP-mRNA through the KH III-IV domain. Febs J 273:5678-5690.

Peltz SW, Brown AH, Jacobson A. 1993. mRNA destabilization triggered by premature translational termination depends on at least three cis-acting sequence elements and one trans-acting factor. Genes Dev 7:1737-1754.

Rehwinkel J, Behm-Ansmant I, Gatfield D, Izaurralde E. 2005. A crucial role for GW182 and the DCP1:DCP2 decapping complex in miRNA-mediated gene silencing. RNA 11:1640-1647.

References

Rehwinkel J, Letunic I, Raes J, Bork P, Izaurralde E. 2005. Nonsense-mediated mRNA decay factors act in concert to regulate common mRNA targets. RNA 11:1530-1544.

Reichenbach P, Hoss M, Azzalin CM, Nabholz M, Bucher P, Lingner J. 2003. A human homolog of yeast Est1 associates with telomerase and uncaps chromosome ends when overexpressed. Curr Biol 13:568-574.

Resta N, Susca FC, Di Giacomo MC, Stella A, Bukvic N, Bagnulo R, Simone C, Guanti G. 2006. A homozygous frameshift mutation in the ESCO2 gene: evidence of intertissue and interindividual variation in Nmd efficiency. J Cell Physiol 209:67-73.

Roy SW. 2006. Intron-rich ancestors. Trends Genet 22:468-471.

Roy SW, Gilbert W. 2006. The evolution of spliceosomal introns: patterns, puzzles and progress. Nat Rev Genet 7:211-221.

Ruiz-Echevarria MJ, Gonzalez CI, Peltz SW. 1998. Identifying the right stop: determining how the surveillance complex recognizes and degrades an aberrant mRNA. Embo J 17:575-589.

Ruiz-Echevarria MJ, Peltz SW. 2000. The RNA binding protein Pub1 modulates the stability of transcripts containing upstream open reading frames. Cell 101:741-751.

Schwartz AM, Komarova TV, Skulachev MV, Zvereva AS, Dorokhov YL, Atabekov JG. 2006. Stability of Plant mRNAs Depends on the Length of the 3 -Untranslated Region. Biochemistry (Mosc) 71:1377-1384.

Serin G, Gersappe A, Black JD, Aronoff R, Maquat LE. 2001. Identification and characterization of human orthologues to Saccharomyces cerevisiae Upf2 protein and Upf3 protein (Caenorhabditis elegans SMG-4). Mol Cell Biol 21:209-223.

Sheth U, Parker R. 2003. Decapping and decay of messenger RNA occur in cytoplasmic processing bodies. Science 300:805-808.

Sheth U, Parker R. 2006. Targeting of aberrant mRNAs to cytoplasmic processing bodies. Cell 125:1095-1109.

Shibuya T, Tange TO, Sonenberg N, Moore MJ. 2004. eIF4AIII binds spliced mRNA in the exon junction complex and is essential for nonsense-mediated decay. Nat Struct Mol Biol 11:346-351.

Shyu AB, Wilkinson MF, van Hoof A. 2008. Messenger RNA regulation: to translate or to degrade. EMBO J 27:471-481.

Silva AL, Ribeiro P, Inacio A, Liebhaber SA, Romao L. 2008. Proximity of the poly(A)-binding protein to a premature termination codon inhibits mammalian nonsense-mediated mRNA decay. RNA 14:563-576.

Singh G, Indrani R, Lykke-Andersen J. 2008. A competition between stimulators and antagonists of Upf complex recruitment governs human nonsense-mediated mRNA decay. PLoS Biol 6:e111.

Singh G, Jakob S, Kleedehn MG, Lykke-Andersen J. 2007. Communication with the exon-junction complex and activation of nonsense-mediated decay by human upf proteins occur in the cytoplasm. Mol Cell 27:780-792.

Snow BE, Erdmann N, Cruickshank J, Goldman H, Gill RM, Robinson MO, Harrington L. 2003. Functional conservation of the telomerase protein Est1p in humans. Curr Biol 13:698-704.

Sun X, Li X, Moriarty PM, Henics T, LaDuca JP, Maquat LE. 2001. Nonsense-mediated decay of mRNA for the selenoprotein phospholipid hydroperoxide glutathione peroxidase is detectable in cultured cells but masked or inhibited in rat tissues. Mol Biol Cell 12:1009-1017.

Tange TO, Nott A, Moore MJ. 2004. The ever-increasing complexities of the exon junction complex. Curr Opin Cell Biol 16:279-284.

Thein SL, Hesketh C, Taylor P, Temperley IJ, Hutchinson RM, Old JM, Wood WG, Clegg JB, Weatherall DJ. 1990. Molecular basis for dominantly inherited inclusion body beta-thalassemia. Proc Natl Acad Sci U S A 87:3924-3928.

Unterholzner L, Izaurralde E. 2004. SMG7 acts as a molecular link between mRNA surveillance and mRNA decay. Mol Cell 16:587-596.

Usuki F, Yamashita A, Higuchi I, Ohnishi T, Shiraishi T, Osame M, Ohno S. 2004. Inhibition of nonsense-mediated mRNA decay rescues the phenotype in Ullrich's disease. Ann Neurol 55:740-744.

van Hoof A, Frischmeyer PA, Dietz HC, Parker R. 2002. Exosome-mediated recognition and degradation of mRNAs lacking a termination codon. Science 295:2262-2264.

van Nimwegen E. 2003. Scaling laws in the functional content of genomes. Trends Genet 19:479-484.

Viegas MH, Gehring NH, Breit S, Hentze MW, Kulozik AE. 2007. The abundance of RNPS1, a protein component of the exon junction complex, can determine the variability in efficiency of the Nonsense Mediated Decay pathway. Nucleic Acids Res 35:4542-4551.

References

Vikesaa J, Hansen TV, Jonson L, Borup R, Wewer UM, Christiansen J, Nielsen FC. 2006. RNA-binding IMPs promote cell adhesion and invadopodia formation. Embo J 25:1456-1468.

Vogel J, Wagner EG. 2007. Target identification of small noncoding RNAs in bacteria. Curr Opin Microbiol 10:262-270.

Wang W, Czaplinski K, Rao Y, Peltz SW. 2001. The role of Upf proteins in modulating the translation read-through of nonsense-containing transcripts. EMBO J 20:880-890.

Wang X, Kiledjian M, Weiss IM, Liebhaber SA. 1995. Detection and characterization of a 3' untranslated region ribonucleoprotein complex associated with human alpha-globin mRNA stability. Mol Cell Biol 15:1769-1777.

Weischenfeldt J, Damgaard I, Bryder D, Theilgaard-Monch K, Thoren LA, Nielsen FC, Jacobsen SE, Nerlov C, Porse BT. 2008. NMD is essential for hematopoietic stem and progenitor cells and for eliminating by-products of programmed DNA rearrangements. Genes Dev 22:1381-1396.

Welch EM, Barton ER, Zhuo J, Tomizawa Y, Friesen WJ, Trifillis P, Paushkin S, Patel M, Trotta CR, Hwang S, Wilde RG, Karp G, Takasugi J, Chen G, Jones S, Ren H, Moon YC, Corson D, Turpoff AA, Campbell JA, Conn MM, Khan A, Almstead NG, Hedrick J, Mollin A, Risher N, Weetall M, Yeh S, Branstrom AA, Colacino JM, Babiak J, Ju WD, Hirawat S, Northcutt VJ, Miller LL, Spatrick P, He F, Kawana M, Feng H, Jacobson A, Peltz SW, Sweeney HL. 2007. PTC124 targets genetic disorders caused by nonsense mutations. Nature.

Wells SE, Hillner PE, Vale RD, Sachs AB. 1998. Circularization of mRNA by eukaryotic translation initiation factors. Mol Cell 2:135-140.

Weng Y, Czaplinski K, Peltz SW. 1996. Identification and characterization of mutations in the UPF1 gene that affect nonsense suppression and the formation of the Upf protein complex but not mRNA turnover. Mol Cell Biol 16:5491-5506.

Wilschanski M, Yahav Y, Yaacov Y, Blau H, Bentur L, Rivlin J, Aviram M, Bdolah-Abram T, Bebok Z, Shushi L, Kerem B, Kerem E. 2003. Gentamicin-induced correction of CFTR function in patients with cystic fibrosis and CFTR stop mutations. N Engl J Med 349:1433-1441.

Wittmann J, Hol EM, Jack HM. 2006. hUPF2 silencing identifies physiologic substrates of mammalian nonsense-mediated mRNA decay. Mol Cell Biol 26:1272-1287.

Yamashita A, Ohnishi T, Kashima I, Taya Y, Ohno S. 2001. Human SMG-1, a novel phosphatidylinositol 3-kinase-related protein kinase, associates with components of the mRNA surveillance complex and is involved in the regulation of nonsense-mediated mRNA decay. Genes Dev 15:2215-2228.

Yang Z, Jakymiw A, Wood MR, Eystathioy T, Rubin RL, Fritzler MJ, Chan EKL. 2004. GW182 is critical for the stability of GW bodies expressed during the cell cycle and cell proliferation. J Cell Sci 117:5567-5578.

Yoine M, Nishii T, Nakamura K. 2006. Arabidopsis UPF1 RNA helicase for nonsense-mediated mRNA decay is involved in seed size control and is essential for growth. Plant Cell Physiol 47:572-580.

6. Appendix

6.1. Summary

In order to assure the high accuracy of gene expression, eukaryotes have evolved diverse quality-control mechanisms that recognize and degrade mRNAs that have not completed nuclear pre-mRNA processing or that fail to encode a complete and functional protein. One of the best studied surveillance mechanisms is nonsense-mediated mRNA decay (NMD) which recognizes and degrades mRNAs harboring a premature termination codon (PTC). Moreover, depletion of the core NMD factor UPF1 alters the steady-state level of about 10 % of all human mRNAs, indicating that NMD also contributes to the posttranscriptional regulation of many physiological mRNAs.

Yet, the molecular mechanism of NMD is not solved. A central step in the mechanism of NMD is the distinction of a PTC from a normal translation termination codon. Clearly, translation is a requisite for NMD to occur. Otherwise, the proposed models for PTC recognition in the different species studied so far are remarkably different, but they essentially fall into two broad categories. The "downstream marker model" proposes that "marker proteins", deposited downstream of a PTC, distinguish a PTC from a normal stop codon. This model has led to the suggestion that a termination codon located upstream of the marker proteins will be recognized as premature, which ultimately leads to the degradation of the mRNA. This model also predicts that the marker proteins bound to the mRNA are displaced during a first round of translation, therefore rendering this mRNA immune to NMD for the subsequent rounds of translation. With the identification of the exon junction complex (EJC) as a downstream marker for mammalian NMD, this model has become the dominating model for mammalian NMD during the last years. The alternative model for PTC recognition invokes a "*faux* 3' untranslated region" (3'UTR). This model postulates that proper (or efficient) translation termination requires a termination-promoting signal, and that the absence of this signal typifies aberrant translation termination at a PTC, which in turn leads to degradation of the mRNA. This model was mainly based on studies in *S. cerevisiae*, but a part of my thesis contributes to the discovery that the *faux* 3'UTR model is partially also true for mammalian PTC recognition. We found that normal mRNAs can be turned into an NMD substrate by elongating the 3'UTR in mammalian cells. Furthermore, PTC-containing mRNAs can be stabilized by bringing the poly(A) tail in close spatial proximity of the PTC, and consistently, PTC-containing mRNAs are stablizied by tethering of poly(A)-binding protein. Together with studies performed in *C. elegans, D. melanogaster, A. thaliana* and humans during the last years, we propose a unified model for PTC recognition, which essentially extends the *faux* 3'UTR model to all eukaryotes studied so far and which introduces downstream markers (EJC in mammalian cells, downstream sequence elements in *S. cerevisiae* (DSE)) as enhancers of NMD. Most importantly, these results demonstrate that spatial rearrangements of the 3'UTR can modulate the NMD pathway and thereby provide a novel mechanism for posttranscriptional gene regulation.

The mechanism by which the PTC-containing mRNA is degraded and the subcellular localization of NMD are still unclear. In *S. cerevisiae*, PTC+ mRNAs were found to localize to so called cytoplasmic processing bodies (P-bodies). In contrast, P-bodies are not essential for NMD in *D. melanogaster*. I found that P-bodies are not required for mammalian NMD, although NMD factors co-localize with P-bodies in mammalian cells.

Another important aspect is the development of therapeutic approaches for NMD-related diseases, as NMD modulates the phenotype of numerous genetic diseases; it has been estimated that one third of all disease-causing mutations generate a PTC. Based on our unified model for PTC recognition, we have designed a potential therapeutic approach to stabilize mRNAs bearing mutations in the 3'UTR that have the potential to alter the spatial relationship between the termination codon and the poly(A) tail. I could show that a PTC-containing mRNA can be stabilized in cell culture using this approach, but I could not yet test it on a disease-relevant allele.

Beside the investigation of NMD, I also worked with another quality-control mechanism. It was discovered in our lab that some PTCs can also lead to transcriptional silencing of its cognate gene via chromatin remodeling by a mechanism termed nonsense-mediated transcriptional gene silencing (NMTGS). I could show that NMTGS depends on the translation of the PTC+ mRNA and that NMTGS can be reversed by RNAi-mediated depletion of the NMD core-factor UPF1. The result that translation and UPF1 are required for NMTGS suggests that NMD and NMTGS are linked mechanistically and it provides the first direct link between the transcriptional silencing and the machinery of PTC recognition.

6.2. Acknowledgments

Zuerst möchte ich Oliver Mühlemann danken. Mit seiner enthusiastischen Art war er für mich ein motivierender Betreuer. Ich konnte viel von seinem Wissen und seiner Erfahrung profitieren, und ich schätzte es sehr, dass er sich immer Zeit für Diskussionen genommen hat und immer offen für neue Ideen war. Ich wünsche ihm und seinem wachsenden Team viel Glück und Erfolg für die Zukunft!

Speziell danken möchte ich auch Andrea Eberle und Marc Ruepp. Ich habe es genossen mit euch zusammen zu arbeiten, Meetings zu besuchen, Kaffee zu trinken und Feste zu feiern. Ich wünsche euch weiterhin viel Glück und Erfolg!

Weiter möchte ich Karin, Nicole, Reto, Raphael, Barbara, Rodolfo, Maria, Larissa, Hasmik, Kathrin, und Judith für die schöne Zeit im und ausserhalb des Labors danken!

Vielen Dank auch an Witold Filipowicz für seine Funktion als Co-Referee und an Ulrich Baumann für sein Mentoring.

Nicht zu vergessen ist der grosse Einsatz von Myriam, Diana und Andy. Ohne euch wäre es unmöglich im Labor zu arbeiten!

Ich möchte mich auch bei meinen Eltern, bei meinem Bruder und meinen Freunden in Bern bedanken. Ohne euch hätte ich nicht die Kraft und Energie gehabt, mich während dieser langen Ausbildungszeit immer wieder zu motivieren und mit Freude an die Arbeit zu gehen!

6.3.1 Paper I

Lukas Stalder, Andrea Eberle and Oliver Mühlemann. Quality control of gene expression: recognition of nonsense mRNAs prevents synthesis of truncated proteins. BioForum Europe, (2007) 11:21-23.

Quality Control of Gene Expression
Recognition of Nonsense mRNAs Prevents Synthesis of Truncated Proteins

To ensure accuracy of gene expression, cells have evolved quality control mechanisms to monitor the different steps of gene expression. One of the most intensively studied surveillance mechanisms is called nonsense-mediated mRNA decay (NMD), a process that recognizes and degrades transcripts containing a premature translation-termination codon (PTC). Thus, NMD prevents the cell from production of truncated proteins, which may act in a dominant-negative way. Several questions in NMD research have attracted much attention in recent years: How is a PTC recognized and discriminated from a physiological stop codon, and by which mechanism does a PTC trigger fast mRNA degradation? What is the role of NMD in the regulation of gene expression and what is the effect of NMD in genetic diseases?

Oliver Mühlemann, Ph.D., Research Group Leader; Andrea Eberle, PhD Student; Lukas Stalder, PhD Student, University of Bern

Various Processes Generate PTCs

PTCs can emerge from frameshift or nonsense mutations in the genome, as well as from errors in every step of gene expression, from the transcription of the nascent mRNA to its translation. The production of a correct mRNA requires a sequence of complicated biochemical processes in the nucleus: transcription, capping, splicing, 3' end formation, and the export to the cytoplasm (fig. 1A). The error rate during transcription is generally low and therefore only seldom a PTC is generated. In contrast, PTCs are frequently acquired by alternative splicing. It was estimated that in mammals, approximately one third of the alternatively spliced transcripts contain PTCs and are substrates for NMD [1]. This is a large fraction of the total mRNA population, given that about 74% of human multi exon genes are alternatively spliced. T cell receptor and immunoglobulin genes represent a special class of genes where PTCs are commonly acquired as a result of programmed V(D)J rearrangements during lymphocyte development [2].

CELL BIOLOGY

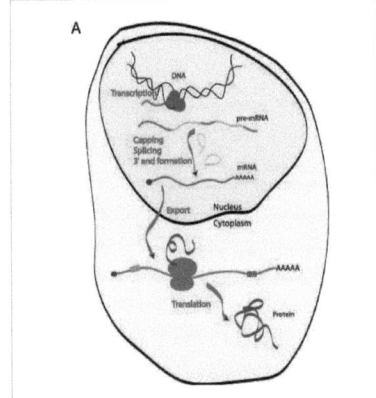
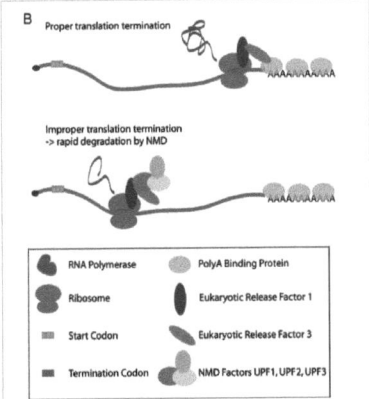

Fig. 1: (A) The general pathway of gene expression. DNA is transcribed into pre-mRNA by RNA polymerase. The pre-mRNA maturates in the cell nucleus: the 5' end is modified (capping), introns are removed by splicing, and the 3' end is cleaved and polyadenylated. The mature mRNA is then exported to the cytoplasm, where it is translated into protein by ribosomes.
(B) Proper versus improper translation termination. When a ribosome stalls at a termination codon, the eukaryotic release factor 1 and 3 bind instead of an aminoacetylated tRNA. Proper/efficient translation termination requires the interaction of the terminating ribosome with the polyA-binding protein. In contrast, our NMD model postulates that failure of the polyA-binding protein to interact with the terminating ribosome is a hallmark of improper/inefficient translation termination and leads to the recruitment of the UPF1-3 factors and to subsequent rapid degradation of the mRNA.

The Mechanism of NMD

NMD is active in all eukaryotes investigated so far, and the three core factors UPF1, UPF2 and UPF3 are conserved from yeast to human. NMD in higher eukaryotes involves additional factors that are not present in yeast. According to the prevailing model for mammalian NMD, a termination codon is identified as premature by its position relative to the last exon-exon junction. This model postulates that transcripts harboring a termination codon more than 50-55 nucleotides upstream of the last exon-exon junction will be subjected to NMD [3]. But recent results suggest that this model needs revision, because a fair number of PTC-containing transcripts violate this rule. In one of those cases that our laboratory investigated, we found that the distance between the termination codon and the poly(A) tail is an important parameter for triggering NMD in human cells, which fits well with data from yeast studies [4, 5]. Therefore, we propose an evolutionary conserved mechanism for PTC recognition, by which a termination codon is recognized as premature, if the ribosome stalls at the termination codon where it fails to interact with factors bound to the poly(A) tail that are necessary for proper translation termination. According to this revised NMD model, the binding of the NMD factors to the mRNA and its subsequent degradation would be the consequence of improper/inefficient translation termination (fig. 1B).

The Role of NMD in Gene Expression

NMD not only rids the cell of mRNAs coding for truncated proteins, but also regulates the expression of ~10% of the transcriptome in yeast, *Drosophila* and human cells [5]. Among the transcripts subjected to NMD in human cells, several groups of genes coding for functionally related proteins were found, including translation factors and ribosomal proteins [1]. These observations suggest that NMD contributes to control whole metabolic pathways by regulating the expression of probably only few mRNAs. NMD is not essential for yeast under laboratory conditions, but the knockout of the NMD factor UPF1 in mouse embryos is lethal, suggesting that NMD might be essential in mammals [6]. Additional to the function in NMD, some NMD factors have been reported to function also in telomere maintenance, transcription and translation regulation, RNA interference, cell proliferation, cell cycle control, cellular transport and organization, and metabolism. NMD may also control noncoding RNAs, which are especially important in higher eukarotes, where a high percentage of the genome is transcribed into noncoding RNAs. These examples demonstrate that in addition to its role as a quality control system, NMD also contributes to the regulation of gene expression.

NMD in Human Diseases

It is estimated that up to 30% of all alleles contributing to genetic diseases in humans harbour a PTC [7]. In these diseases, NMD modulates the phenotype in a beneficial or harmful way. In the beneficial cases, NMD prevents the accumulation of truncated proteins that would otherwise act in a dominant-negative way and disturb cell functions. However, NMD can also cause more severe clinical symptoms by inhibiting the synthesis of truncated proteins that, if expressed, would still be functional. Therefore, the exact phenotype of a disease often depends on the NMD efficiency, which in turn is mainly determined by the location of the PTC, and on the properties of the truncated protein (table 1).

Beneficial Effects of NMD

β-thalassemias represent a well-documented example of a human disease where NMD is beneficial. Patients suffering from β-thalassemias have a nonsense

CELL BIOLOGY

Table 1: Examples of genetic diseases where NMD modulates the phenotype (adapted from [8])

Gene name	NMD efficiency	Effect of Mutation / Phenotype
NMD beneficial		
β-globin	high	heterozygotes healthy; recessively inherited β-thalassemia major
	low	dominantly inherited β-thalassemia intermedia
Rhodopsin	high	heterozygotes have abnormalities on retinogram, but no clinical disease; recessively inherited blindness
	low	dominantly inherited blindness
Receptor tyrosine kinase -like orphan receptor 2	high	heterozygotes healthy; recessively inherited Robinow syndrome (orodental abnormalities, hypoplastic genitalia, multiple rib/vertebral anomalies)
	low	dominantly inherited brachydactyly type B (shortening of digits and metacarpals)
Cone-rod homeobox	high	mutation found in unaffected heterozygotes (no homozygotes found)
	low	dominantly inherited retinal disease
Coagulation factor X	high	heterozygotes healthy; recessively inherited bleeding tendency
	low	dominantly inherited bleeding tendency
Interferon γ receptor 1	high	heterozygotes healthy; recessively inherited susceptibility to mycobacterial infection
	low	dominantly inherited susceptibility to mycobacterial infection
von Willebrand factor	high	heterozygotes healthy; recessively inherited type 3 von Willebrand disease
	low	dominantly inherited type 2A disease
NMD detrimental		
Dystrophin	high	severe form of muscular dystrophy (Duchenne muscular dystrophy)
	low	milder form of muscular dystrophy (Becker muscular dystrophy)
CFTR (cystic fibrosis)	high	severe form of cystic fibrosis
	low	milder form of cystic fibrosis

mutation in the β-globin gene. In patients homozygote for the mutation, the amount of β-globin is insufficient and they develop a severe anaemia. In contrast, heterozygote carriers can synthesize enough β-globin from the functional allele and are asymptomatic in most cases. In these individuals, the nonsense transcript of the mutated allele is degraded by NMD, and almost no truncated β-globin protein is produced. A variant of β-thalassemia, where the PTC does not trigger degradation of the mutated mRNA efficiently, demonstrates the importance of NMD: the synthesized truncated β-globin forms toxic inclusion bodies in the cells, which leads to a severe phenotype even in heterozygotic individuals [8].

Harmful Effects of NMD

But there are also several diseases where NMD causes a more severe phenotype, like cystic fibrosis or muscular dystrophy. In these cases, the PTC-containing transcript encodes a truncated protein with some residual function, but NMD leads to a dramatic reduction of these protein levels and thereby worsens the disease phenotype. The severity of dystrophy correlates with the efficiency by which a PTC-containing mRNA is recognized and degraded. If located towards the 3' end of the gene, the transcript is a poor substrate for NMD and these mutations lead to a milder phenotype, called Becker muscular dystrophy. In the more severe form (Duchenne muscular dystrophy), the PTC is located further upstream in the gene, where it triggers more efficient NMD [7, 8].

Clinical Approaches

Different clinical approaches are currently under investigation to treat NMD-related diseases. In the cases where NMD is responsible for worse clinical symptoms, the strategy is to inhibit NMD by reducing the efficient translation termination with antibiotics or suppressor tRNAs, which lead to a read-through at termination codons [9]. Despite of the expected side effects of these approaches, several antibiotics are in clinical trials and a lot of effort is undertaken to find new substances to reduce their toxic side effects in humans. As an alternative, gene therapy approaches are used to repair nonsense mutations already in the genome [9]. And in cases where aberrant splicing produces nonsense mRNAs, gene therapy techniques that correct the splicing pattern are promising strategies.

Concluding Remarks

Until now, the molecular mechanism used by the cell to distinguish PTCs from "normal" termination codons is poorly understood and currently the focus of intensive research in several laboratories. Likewise, we are only beginning to understand the molecular details of the effector pathways leading to mRNA degradation and even less is known about the role of NMD in general regulation of gene expression. On the other hand, because of the documented role of NMD as an important modulator of the clinical manifestations of many genetic diseases, a better understanding of the molecular mechanisms underlying NMD is the key for developing techniques that allow regulation of NMD efficiency of specific transcripts and therewith more specific therapies for NMD related diseases.

References

[1] Lewis et al.: Proc. Nat. Acad. Sci. U.S.A. 100, 189-192 (2003)
[2] Li and Wilkinson: Immunity 8, 135-141 (1998)
[3] Lejeune and Maquat: Curr. Opin. Cell. Biol. 17, 309-315 (2005)
[4] Bühler et al.: Nat. Struct. Mol. Biol. 13, 462-464 (2006)
[5] Amrani et al.: Nat. Rev. Mol. Cell. Biol. 7, 415-425 (2006)
[6] Medghalchi et al.: Hum. Mol. Genet. 10, 99-105 (2001)
[7] Mendell and Dietz: Cell 107, 411-414 (2001)
[8] Holbrook et al.: Nat. Genet. 36, 801-808 (2004)
[9] Keeling et al.: in Nonsense-mediated mRNA decay (Ed. L.E. Maquat), Landes Bioscience, Georgetown, TX, U.S.A. (2006)

The authors wish to thank Ebbe Sloth Andersen for his cartoon.

▶ www.eMagazineBIOforum.com

CONTACT:
Andrea Eberle
Lukas Stalder
Oliver Mühlemann, PhD
Institute of Cell Biology
University of Bern, Switzerland
Tel.: +41 31 631 4627
Fax: +41 31 631 4616
oliver.muehlemann@izb.unibe.ch
www.izb.unibe.ch/res/muehlemann/index.php

6.3.2. Paper II

Stalder L, Mühlemann O. Transcriptional silencing of nonsense codon-containing immunoglobulin µ genes requires translation of its mRNA. J Biol Chem. 2007 Jun 1;282(22):16079-85

This publication can not be printed here due to copyright protection. Please download this publication from www.jbc.org

6.3.3. Paper III

Eberle AB[1], Stalder L[1], Mathys H, Orozco RZ, Mühlemann O. Posttranscriptional gene regulation by spatial rearrangement of the 3' untranslated region. PLoS Biol. 2008 Apr 29;6(4):e92.

[1]equally contributed

Appendix: Paper III

Posttranscriptional Gene Regulation by Spatial Rearrangement of the 3' Untranslated Region

Andrea B. Eberle[o], Lukas Stalder[o], Hansruedi Mathys[¤], Rodolfo Zamudio Orozco, Oliver Mühlemann*

Institute of Cell Biology, University of Berne, Berne, Switzerland

Translation termination at premature termination codons (PTCs) triggers degradation of the aberrant mRNA, but the mechanism by which a termination event is defined as premature is still unclear. Here we show that the physical distance between the termination codon and the poly(A)-binding protein PABPC1 is a crucial determinant for PTC recognition in human cells. "Normal" termination codons can trigger nonsense-mediated mRNA decay (NMD) when this distance is extended; and vice versa, NMD can be suppressed by folding the poly(A) tail into proximity of a PTC or by tethering of PABPC1 nearby a PTC, indicating an evolutionarily conserved function of PABPC1 in promoting correct translation termination and antagonizing activation of NMD. Most importantly, our results demonstrate that spatial rearrangements of the 3' untranslated region can modulate the NMD pathway and thereby provide a novel mechanism for posttranscriptional gene regulation.

Citation: Eberle AB, Stalder L, Mathys H, Zamudio Orozco R, Mühlemann O (2008) Posttranscriptional gene regulation by spatial rearrangement of the 3' untranslated region. PLoS Biol 6(4): e92. doi:10.1371/journal.pbio.0060092

Introduction

Nonsense-mediated mRNA decay (NMD) represents a translation-dependent posttranscriptional mRNA quality control process that selectively degrades mRNAs containing premature termination codons (PTCs), thereby preventing the synthesis of truncated, potentially deleterious proteins [1,2]. Because one-third of all known disease-causing mutations are predicted to generate a PTC, NMD serves as an important modulator of genetic disease phenotypes in humans [3,4]. Hence, understanding the molecular mechanism of NMD will facilitate the future development of gene-specific therapies for many genetic diseases. Interestingly, NMD affects 3%–10% of the transcriptome of *Saccharomyces cerevisiae*, *Drosophila melanogaster*, and mammals, indicating that NMD, in addition to its quality control function, is also involved in regulating the expression of many physiological transcripts (reviewed in [5]).

The three Upf (*Up-f*rameshift) proteins—Upf1, Upf2, and Upf3—work at the heart of the NMD pathway in all organisms studied. The Upf proteins were first discovered in genetic screens in *S. cerevisiae* and *Caenorhabditis elegans*, and orthologs have subsequently been identified in other eukaryotes (reviewed in [6]). Upf1 is an ATP-dependent RNA helicase, and a mutation in the ATPase domain abolishes its 5'-to-3' helicase activity and its function in NMD [7–9]. Human Upf2 contains three conserved middle of eIF4G-like (MIF4G) domains, multiple putative nuclear localization signals in its N-terminus, and a putative nuclear export signal. Upf2 interacts with Upf1 and Upf3, and the three proteins can be isolated as a complex [10–13]. Upf3 is the least conserved component among the Upf proteins [6]. Humans contain two different UPF3 genes, encoding Upf3a and Upf3b (also known as Upf3X since the corresponding gene maps to the X chromosome, respectively) [11,13]. In addition to Upf1–3, metazoans contain additional NMD factors (Smg1, Smg5, Smg6, and Smg7) that are involved in regulating the phosphorylation state and therewith the activity of Upf1 [14].

While the phenomenon of NMD and its impact on gene expression are well documented, the understanding of the underlying molecular mechanisms is still fragmented. A central question is how PTCs are recognized and discriminated from natural termination codons (TCs). The current models for NMD differ remarkably between mammals and other eukaryotes. While all models agree that translation is required for NMD, studies of mammalian genes indicate that NMD in higher eukaryotes also depends on pre-mRNA splicing. The current model for mammalian NMD postulates that PTC recognition requires an interaction between an exon junction complex (EJC) bound to the mRNA downstream of the TC and the terminating ribosome [15,16]. However, several examples of NMD in mammals have been reported that are inconsistent with this EJC-dependent NMD model ([17] and references therein). Recently, we showed that PTCs in the terminal exon of Ig-μ minigenes (miniμ) elicit NMD dependent on the length of their 3' untranslated region

Academic Editor: Marv Wickens, University of Wisconsin, United States of America

Received November 13, 2007; Accepted March 4, 2008; Published April 29, 2008

Copyright: © 2008 Eberle et al. This is an open-access article distributed under the terms of the Creative Commons Attribution License, which permits unrestricted use, distribution, and reproduction in any medium, provided the original author and source are credited.

Abbreviations: EJC, exon junction complex; FB, foldback; NFB, no foldback; NMD, nonsense-mediated mRNA decay; PABP, poly(A)-binding protein; PTC, premature termination codon; RT-PCR, reverse-transcriptase PCR; RT-qPCR, reverse-transcriptase quantitative PCR; TC, termination codon; UTR, untranslated region

* To whom correspondence should be addressed. E-mail: oliver.muehlemann@izb.unibe.ch

[o] These authors contributed equally to this work.

[¤] Current address: Friedrich Miescher Institute for Biomedical Research, Basel, Switzerland

Author Summary

Correct expression of the genetic information is essential for life, and several quality control systems have evolved to ensure accurate protein synthesis. One of these processes, termed nonsense-mediated mRNA decay (NMD), detects inappropriate termination of mRNA translation at premature termination codons (PTCs) and triggers degradation of the aberrant mRNA. Although the occurrence of NMD is well documented in yeast, worms, flies, mammals, and plants, the mechanism by which a termination event is defined as premature is still unclear, and different models have been proposed for different species. For mammals, the current prevailing view is that a termination codon is identified as premature and elicits NMD when it is located upstream of the 3'-most exon junction complex. However, well-documented examples of NMD triggered by PTCs in the last exon challenge this "mammalian NMD model." Here we show that the physical distance between the termination codon and the poly(A)-binding protein PABPC1 is a crucial determinant for PTC recognition in human cells, indicating an evolutionarily conserved function of PABPC1 in promoting correct translation termination and antagonizing activation of NMD. Most importantly, our results demonstrate that spatial rearrangements of the 3' untranslated region can modulate the NMD pathway and thereby provide a novel, translation-dependent mechanism for posttranscriptional gene regulation.

(UTR) [17]. This is reminiscent of the situation in *D. melanogaster*, *C. elegans*, *S. cerevisiae*, and plants, where PTC recognition occurs independently of splicing and EJC factors, but where instead 3' UTR length and the poly(A)-binding protein (PABP) were found to play an important role [18–20]. The "faux 3' UTR" model, which is based on studies with yeast, postulates that proper translation termination requires an interaction between PABP and eRF3 bound to the terminating ribosome, and that the absence of this positive signal leads to aberrant termination and NMD as a consequence [21]. We report here that our results obtained with human cell lines are consistent with the yeast "faux 3' UTR" model, providing strong evidence for an evolutionarily conserved basic mechanism of PTC recognition in which PABPC1 acts as an NMD antagonizing factor and a role for the mammalian EJC as an NMD enhancer. But most importantly, our data show that mRNA half-lives can be regulated by altering the spatial configuration of their 3' UTRs. This represents a novel, potentially widespread mechanism for posttranscriptional gene regulation by NMD.

Results

3' UTR Extensions Reduce the mRNA Half-Life by NMD

Comparison of relative mRNA levels of miniµ constructs with PTCs at different positions indicated that PTCs located toward the 5' and the 3' ends of the mRNA induce gradually less efficient NMD (Figure 1A). Because mRNAs were shown to adopt a circular conformation that positions the 5' and close to the 3' end by eIF4G bridging the cap-bound factors (eIF4E or CBC) with the poly(A)-binding protein PABPC1 bound to the 3' end [22–24], and based on our previous results [17], we hypothesized that the distance between the TC and the poly(A) tail might be a crucial determinant to identify a TC as premature. Supporting this hypothesis, extension of this distance by insertion of prokaryotic sequence into the 3' UTR redefines the normal TC as NMD-triggering PTC (Figure 1B–1D and [17]). Prokaryotic sequence was chosen for these 3' UTR extensions to minimize the risk of unintentionally inserting binding sites for mammalian RNA-binding proteins that could affect transcript stability. Extending the 3' UTR of a miniµ construct with the full-length coding region (miniµ C3/C4 WT) from 300 to 900 nucleotides reduced the mRNA level to 40%, and an extension to 1,500 nucleotides led to a further reduction to 7% (Figure 1B). Judged from the Northern blot analysis, insertion of these sequences into the 3' UTR did not interfere with pre-mRNA splicing or 3' end formation. We further determined the decay kinetics of the miniµ WT and miniµ WT +1,200 mRNA using the Tet-Off Advanced System in HeLa cells. The 3' UTR extension of 1,200 nucleotides caused a reduction of the mRNA's half-life from 4.16 h (miniµ WT) to 2.16 h (miniµ WT +1,200; Figure 1C). Additionally, the mRNA reduction of miniµ WT+1,200 can be suppressed to variable extent by RNAi-mediated knockdown of Upf1, Upf2, and Upf3b, indicating that it is caused by bona fide NMD (Figures 1D and S1). Depletion of Upf1 resulted in a 16-fold, depletion of Upf2 in a 4-fold, and depletion of Upf3b in an 8-fold increase of miniµWT+1,200 mRNA, respectively. These differences in the extent of NMD suppression most likely reflect different knockdown efficiencies and/or different minimal concentrations of these proteins required to sustain NMD.

A similar extension of the 3' UTR of a β-globin reporter gene from 128 to 705 nucleotides (Figure S2), and extension of the miniµ 3' UTR by a different sequence [17], also caused an Upf1-dependent mRNA reduction, suggesting that the length of the 3' UTR is an important and general determinant to define a TC as premature and trigger NMD independent of the sequence context. Because in all these examples no exon–exon junction and hence no EJC is located downstream of the PTC, and (ii) several lines of evidence suggest that the upstream EJCs have been removed by the first translating ribosome [25], we refer to this form of NMD as "EJC-independent."

Reducing the Physical Distance between the TC and the Poly(A) Tail Suppresses EJC-Independent NMD

Because many transcripts in higher eukaryotes have long 3' UTRs [26,27], simply the number of nucleotides between the TC and the poly(A) tail is unlikely to represent the signal for defining a TC as premature. Instead, we hypothesized that it may be rather the physical distance between the TC and the poly(A) tail that bears the kinetic and regulatory potential to distinguish a proper translation termination event from aberrant termination. According to this model, it should be possible to suppress NMD by reducing this distance. We tested this with so-called "foldback" constructs, in which 26 nucleotides complementary to the sequence located about 50 nucleotides downstream of the PTC were inserted into miniµ immediately upstream of the poly(A) signal. Base pairing of this complementary sequence positions the poly(A) tail in the vicinity of the PTC in these transcripts (Figure 2A). Northern blot analysis (Figure S3A) and reverse transcriptase (RT)-PCR (unpublished data) confirmed that introduction of this intramolecular base pairing did not interfere with splicing or 3' end processing. For this EJC-independent NMD reporter (miniµ C3/H4 ter440 [17]), a mRNA half-life of 2.59 h was observed with the miniµ ter440 "no foldback" (NFB) control construct, whereas the mRNA half-life of the

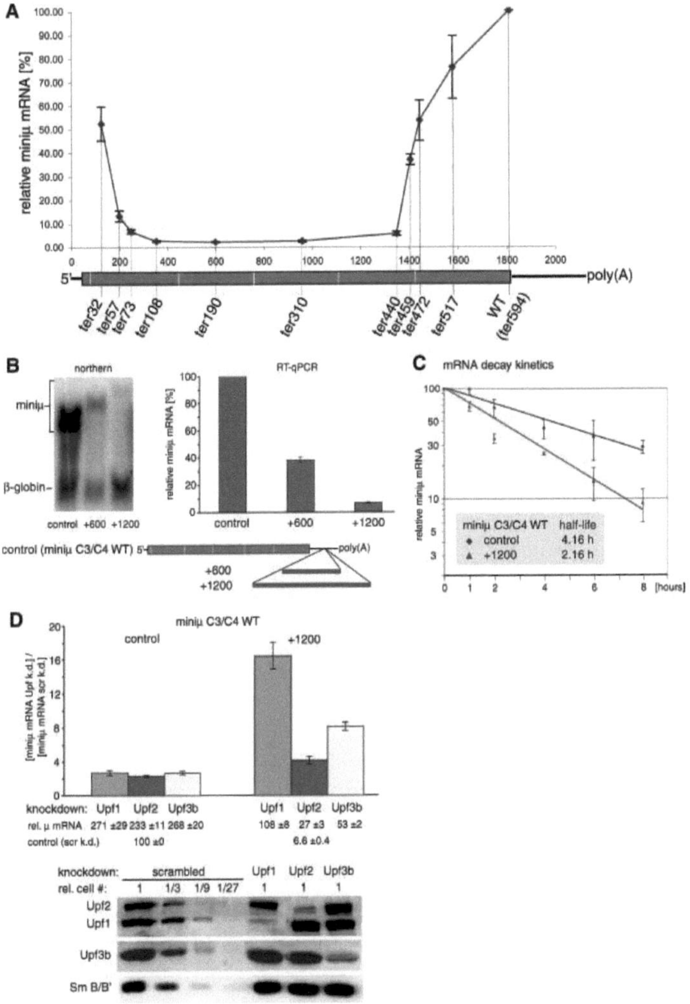
A
relative miniμ mRNA [%]
100.00
90.00
80.00
70.00
60.00
50.00
40.00
30.00
20.00
10.00
0.00
0
200
400
600
800
1000
1200
1400
1600
1800
2000
5'
poly(A)
ter32
ter57
ter73
ter108
ter190
ter310
ter440
ter459
ter472
ter517
WT
(ter594)
B
northern
miniμ
β-globin
control +600 +1200
RT-qPCR
relative miniμ mRNA [%]
100
80
60
40
20
0
control
+600
+1200
control (miniμ C3/C4 WT) 5'
poly(A)
+600
+1200
C
mRNA decay kinetics
relative miniμ mRNA
100
50
30
10
5
3
miniμ C3/C4 WT half-life
control 4.16 h
+1200 2.16 h
0 1 2 4 6 8 [hours]
D
miniμ C3/C4 WT
control
+1200
[miniμ mRNA Upf k.d.] / [miniμ mRNA scr k.d.]
20
16
12
8
4
0
knockdown: Upf1 Upf2 Upf3b
rel. μ mRNA 271 ±29 233 ±11 268 ±20
control (scr k.d.) 100 ±0
Upf1 Upf2 Upf3b
108 ±8 27 ±3 53 ±2
6.6 ±0.4
knockdown: scrambled Upf1 Upf2 Upf3b
rel. cell #: 1 1/3 1/9 1/27 1 1 1
Upf2
Upf1
Upf3b
Sm B/B'

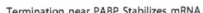

Termination near PABP Stabilizes mRNA

Figure 1. The Distance between the TC and the Poly(A) Tail Is Crucial for NMD
(A) Relative miniμ mRNA levels, normalized to the mRNA levels of cotransfected β-globin WT gene, were determined by RT-qPCR from miniμ constructs with PTCs at the indicated amino acid positions. PTC positions on the mRNA are plotted on the x-axis, and average values and SD of four qPCR runs are shown. The miniμ mRNA is schematically shown below the plot with the positions of the PTCs in the different constructs indicated. In all figures, 5′ and 3′ UTRs are depicted as black lines, ORFs as gray boxes, and exon-exon junctions as vertical white lines.
(B) Northern blot and RT-qPCR analysis of miniμ mRNA with a 3′ UTR extension by 600 or 1,200 nucleotides. RT-qPCR data were obtained and normalized as in (A).
(C) mRNA decay kinetics of the control miniμ C3/C4 WT construct and the +1,200 construct illustrated in (B) were determined in HeLa Tet-Off cells. Relative miniμ mRNAs, normalized to cotransfected β-globin, were analyzed 0, 1, 2, 4, 6, and 8 h after doxycycline addition by RT-qPCR. Average values and SD of two independent experiments with three RT-qPCR runs each are shown.
(D) Effect of RNAi-mediated depletion of Upf1, Upf2, or Upf3b on the relative miniμ mRNA levels of the control and the +1,200 construct. The ratio of the normalized miniμ mRNA levels between the indicated Upf knockdown and the control knockdown (scrambled) is represented. Average and SD of three qPCR runs are shown in (B) and (D). The efficacy of the Upf1, Upf2, and Upf3b knockdown was monitored by Western blotting (lower panel). Detection of SmB/B′ served as loading control.
doi:10.1371/journal.pbio.0060092.g001

corresponding "foldback" (FB) construct was with 7.74 h very similar to the half-lives of the control constructs WT NFB (8.44 h) and WT FB (6.96 h, Figure 2B). Furthermore, the steady-state mRNA level of the miniμ ter440 NFB control construct was reduced to 30% in an Upf1-dependent manner,

whereas the mRNA level of the corresponding FB construct was only marginally reduced compared to the WT and was not affected by RNAi-mediated Upf1 depletion (Figure 2C and 2D). To confirm that the observed NMD suppression of ter440 FB mRNA was specifically dependent on the base

A miniμ C3/H4 foldback and control constructs

B mRNA decay kinetics

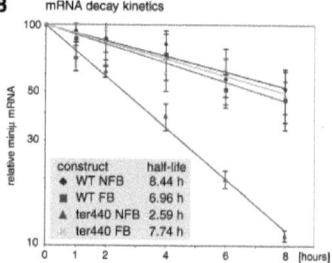

construct	half-life
WT NFB	8.44 h
WT FB	6.96 h
ter440 NFB	2.59 h
ter440 FB	7.74 h

C mRNA increase upon Upf1 knockdown

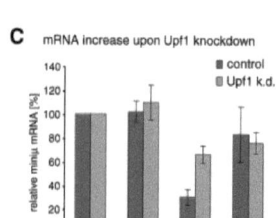

D western blot of Upf1 knockdown

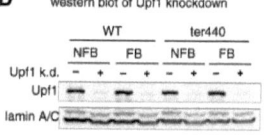

Figure 2. Suppression of EJC-Independent NMD by Poly(A) Tail FB
(A) Schematic illustration of the mRNAs expressed by the indicated constructs. The 26-nucleotide sequence located 42 nucleotides downstream of codon 440 is depicted in red, and the insertion upstream of the poly(A) tail of this sequence (red) or of the complementary sequence (green) is indicated. WT = construct with full-length ORF; ter440 = construct with PTC at codon 440.
(B) Half-lives of the FB mRNAs were measured as described in Figure 1C.
(C) Relative miniμ mRNA levels from the EJC-independent FB constructs shown in (A) normalized to β-globin WT mRNA from a cotransfected plasmid, were measured by RT-qPCR from Upf1-depleted cells (light gray bars) or from control cells expressing a scrambled shRNA (dark gray bars).
(D) The efficacy of the Upf1 knockdown was assessed by Western blotting. Detection of lamin A/C served as loading control.
doi:10.1371/journal.pbio.0060092.g002

Termination near PABP Stabilizes mRNA

ter440 mutFB shows that this mRNA is a substrate for NMD (Figure S4B and S4C). This indicates that NMD suppression of ter440 FB requires the actual formation of the predicted intramolecular base pairs, because abolishing of this base pairing potential renders the transcript NMD-sensitive. Collectively, these results demonstrate that folding the poly(A) tail into the vicinity of a PTC in the terminal exon suppresses EJC-independent NMD.

mRNA Stability of FB Constructs Gradually Decreases with Increasing Distance between TC and Poly(A) Tail

Next we wanted to determine up to which maximal distance from the TC the poly(A) tail is able to suppress NMD. We generated additional FB constructs analogous to miniμ C3/H4 ter440 FB (Figure 2) by inserting different sequences into the poly(A) signal proximal SpeI restriction site that are complementary to different regions downstream of ter440 (Figure 3A). The complementary sequences are between 20 and 30 nucleotides long and designed to have a melting temperature of about 60 °C when base pairing to their target sequence (see Materials and Methods). In HeLa cells transiently transfected with these pTRE-tight FB reporter constructs, we stopped transcription by addition of doxycycline and analyzed the decay kinetics of the FB transcripts. These experiments revealed a gradual destabilization of the mRNAs with increasing distance of the poly(A) tail from the PTC (Figure 3B and 3C). Noteworthy, the two independently determined half-lives for miniμ C3/H4 ter440 FB mRNA differ by less than 1% (7.74 h in Figure 2B and 7.81 h in Figure 3B), indicating that these half-life measurements are highly reproducible and precise. Whereas the half-life of FB mRNA is similar to the half-lives of WT NFB and WT FB (Figure 2B), indicating a complete suppression of NMD, the half-life of FB5 (2.54 h, Figure 3B) is similar to the half-life of ter440 NFB (2.59 h, Figure 2B), indicating a complete loss of NMD suppression. The half-lives of FB2, FB3, and FB4 mRNA fall in-between and indicate a partial loss of NMD suppression. From these results, we conclude that the poly(A) tail-mediated NMD-suppressing activity functions in a distance-dependent manner, as manifested by the gradual decrease of mRNA stability upon increasing distance between TC and poly(A) tail (Figure 3C).

Reducing the Physical Distance between the TC and the Poly(A) Tail Also Suppresses EJC-Enhanced NMD

Our previous results suggested that in mammals, the EJC has adopted a NMD-enhancing function when present downstream of a TC, presumably by increasing the local concentration of the two essential NMD factors Upf2 and Upf3b [17]. We therefore tested if such EJC-enhanced NMD of a miniμ mRNA can also be suppressed by folding back the poly(A) tail (Figures 4 and S5). To this end, we generated a FB construct with a PTC located in the fourth of six exons (miniμ ter310 FB, Figure 4A). After splicing, the mRNA of this minigene construct is expected to harbor two EJCs downstream of the PTC, and the corresponding NFB control mRNA (ter310 NFB) should therefore be subject to efficient EJC-enhanced NMD. Indeed, the mRNA half-life of control construct ter310 NFB was only 1.92 h (Figure 4B) and the steady-state mRNA level only 1.1% of the corresponding WT NFB mRNA (Figure 4C), indicative of efficient EJC-enhanced NMD. In contrast, the mRNA of the PTC-containing FB

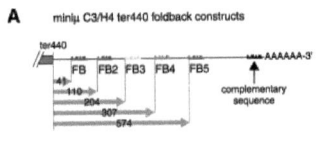

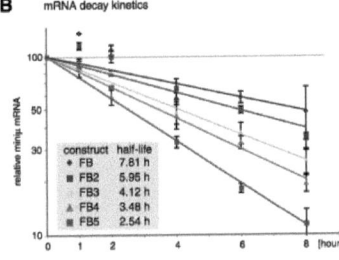

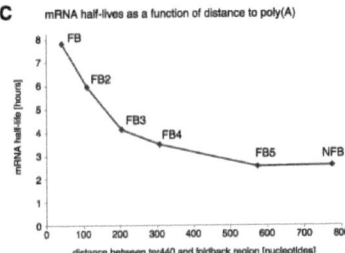

Figure 3. The Distance between the PTC and the Poly(A) Tail Is an Important Criterion for mRNA Stability
(A) Schematic illustration of the miniμ C3/H4 ter440 FB mRNAs expressed by the indicated constructs. Complementary sequences to the regions FB, FB2, FB3, FB4, and FB5 were inserted as double-stranded oligonucleotides at the position marked by the arrow. The distance (in number of nucleotides) between the PTC (ter440) and the first base of the respective base pairing region are indicated below.
(B) Half-lives of the FB mRNAs were measured as described in Figure 1C. Average values and SD from one experiment with three RT-qPCR runs are shown.
(C) Half-lives of the indicated mRNAs (data from Figures 2B and 3B) were plotted against the distance between ter440 and the first base of the base-pairing region.
doi:10.1371/journal.pbio.0060092.g003

pairing between the inserted sequence near the poly(A) tail and the complementary region about 50 nucleotides downstream of ter440, we mutated seven nucleotides in this region to abolish the base pairing potential (ter440 mutFB, Figure S4A). The Upf1-dependent mRNA reduction observed with

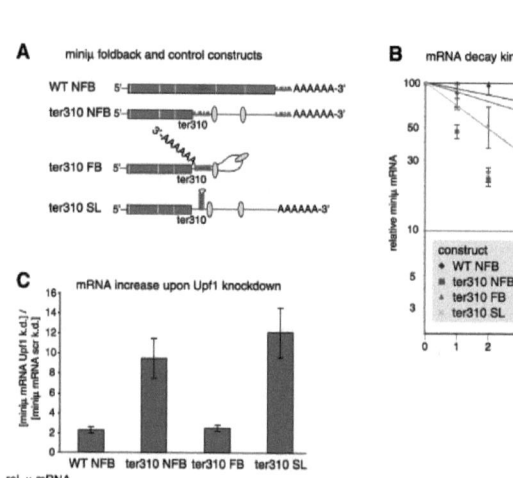

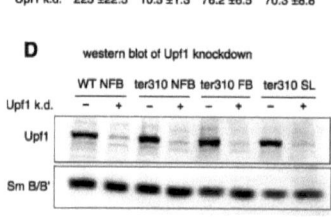

Figure 4. Suppression of EJC-Enhanced NMD by Poly(A) Tail FB
(A) Schematic illustration of the mRNAs expressed by the indicated constructs. The 26-nucleotide sequence located 28 nucleotides downstream of codon 310 is depicted in red, and the insertion upstream of the poly(A) tail of this sequence (red) or of the complementary sequence (green) is indicated. Predicted EJCs in the 3′ UTR are shown by yellow ovals. WT = construct with full-length ORF; ter310 = construct with PTC at codon 310; SL = stemloop control construct with complementary sequence inserted in exon C2.
(B) Half-lives of the FB mRNAs were measured as described in Figure 1C.
(C) Relative miniμ mRNA levels from FB constructs with 3′ UTR introns expressed in cells with (+) or without (−) Upf1 knockdown were measured as in Figure 2C and are shown below the histogram. The histogram depicts the fold increase of miniμ mRNA upon Upf1 knockdown.
(D) The Upf1 knockdown efficacy was assessed by Western blotting as in Figure 2D. Detection of Sm B/B′ served as loading control.
doi:10.1371/journal.pbio.0060092.g004

construct (ter310 FB) was reduced only to 30%–40% of WT NFB, and significantly stabilized, indicated by an average half-life of 4.69 h (Figure 4B). Importantly, the intramolecular base pairing by itself did not stabilize the mRNA when it did not position the poly(A) tail close to the PTC (ter310 SL), and mutations in the base pairing region of ter310 FB reverted the transcript back into an NMD substrate (Figure S4D–S4F). Rather than fitting a simple exponential decay curve, we noticed that ter310 NFB and ter310 SL mRNA exhibit a fast initial decay rate that slows down at lower mRNA levels. The reason for this apparently bi-phasic decay kinetics is currently not known. Because it is not observed with EJC-independent NMD reporters, it may reflect mechanistic differences between EJC-independent and EJC-enhanced NMD (see Discussion). Consistent with the result from the decay assay, NMD inhibition by RNAi-mediated Upf1 depletion (Figure 4C and 4D) or treatment of the cells with the translation inhibitor cycloheximide (Figure S5) elevated the mRNA levels of the NMD substrates ter310 NFB and ter310 SL by a factor of 10–20; ter310 FB behaved like WT

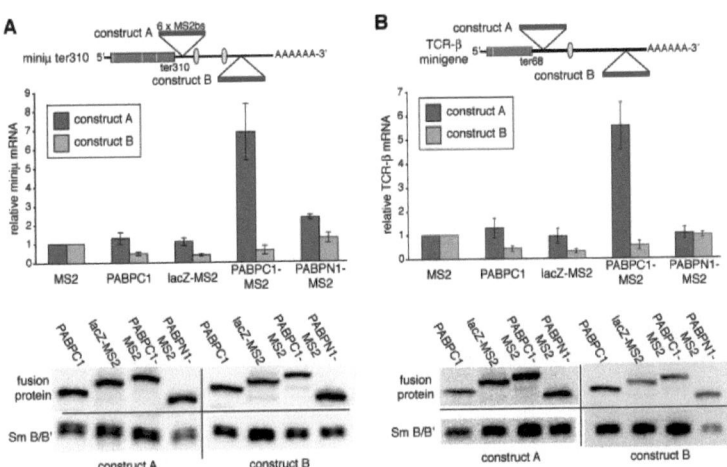

Figure 5. Tethering PABPC1 Nearby a PTC Suppresses NMD
(A and B, upper panels) Shows schematic illustration of PTC-containing reporter genes miniμ (A) and TCRβ (B). A cassette comprising six MS2 binding sites (marked in red) was inserted about 50 nucleotides downstream of a PTC in construct A or further away in construct B. (A and B, middle panels) Relative mRNA levels of construct A (dark bars) and construct B (light bars), normalized to a cotransfected rGPx-1 gene, were measured by RT-qPCR. Average mRNA levels and SD were derived from two independent experiments with two qPCR runs each. (A and B, lower panels) The expression of the fusion proteins were analyzed by immunoblotting using monoclonal mouse α-HA antibody. Detection of SmB/B' served as loading control.
doi:10.1371/journal.pbio.0060092.g005

NFB. We conclude that positioning of the poly(A) tail near a PTC efficiently suppresses both EJC-independent and EJC-enhanced NMD. The latter result shows that the NMD-inhibiting signal (poly(A) tail proximity) efficiently competes with the NMD-promoting signal (the downstream EJC), resulting in a strong attenuation of NMD.

PABPC1 Is an NMD Antagonizing Signal

But which constituent of the poly(A) tail has the capacity to suppress NMD? It was recently shown for *S. cerevisiae* [18] and for *Drosophila* cell lines [19] that the poly(A)-binding protein inhibits NMD when tethered near the PTC on a NMD reporter transcript. To test if tethered poly(A)-binding protein inhibits NMD also in human cells, we tested both the nuclear and the cytoplasmic poly(A)-binding proteins (PABPN1 and PABPC1) expressed as N-terminal fusions to an HA-tagged variant of the MS2 coat protein and assessed their effect on miniμ and TCR-β NMD substrates that harbor six MS2 binding sites either about 50 nucleotides downstream of the PTC (constructs A) or further away as a control (constructs B, Figure 5). Western blotting confirmed comparable expression levels of the different fusion proteins. Tethering of PABPC1 caused a strong increase in both reporter mRNAs when tethered nearby the PTC, indicative for NMD suppression. In contrast, tethering of PABPN1, the MS2 domain alone, PABPC1 expression without the MS2 domain, or a fragment of β-galactosidase with similar mass as PABPC1 did not significantly stabilize the reporter mRNAs. Thus, as in *D. melanogaster*, only the cytoplasmic but not the nuclear PABP inhibited NMD when located in the vicinity of the PTC [19].

Discussion

Evidence for an Evolutionarily Conserved Mechanism for PTC Recognition

In summary, our results demonstrate that in human cells the proximity of PABPC1 provides an important signal for defining a translation termination event as "correct" and prevents degradation of the mRNA by NMD. Vice versa, translation termination too distant from PABPC1 lacks this signal, and as a consequence, NMD ensues. Our results are entirely consistent with models previously proposed for NMD in *S. cerevisiae* [18,21,28] and strongly argue that the basic mechanism for PTC recognition is much more conserved among eukaryotes than previously assumed. In essence, the current data suggest that the two antagonizing signals (PABPC1 proximity and Upf1-3 recruitment, respectively) determine if a translation termination event is defined as premature or correct. The recent characterization of PTC124, a small chemical entity that selectively induces ribosomal readthrough of premature but not normal TCs

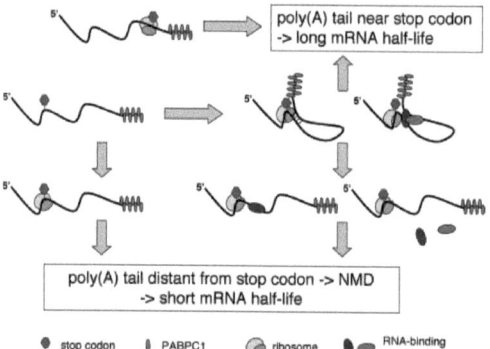

Figure 6. Model for Posttranscriptional Gene Regulation by Spatial Rearrangement of the 3' UTR
Proper translation termination requires a termination-promoting signal from the poly(A) tail. When the ribosome terminates close enough to the poly(A) tail to receive this signal no NMD occurs, and the mRNA remains intact. If the physical distance between the stop codon and the poly(A) tail is too large to allow transmission of the termination-promoting signal, NMD ensues, resulting in a short half-life and low steady-state level of the mRNA. The physical distance between the stop codon and the poly(A) tail depends on the 3-D structure of the 3' UTR. The 3' UTR structure can be reconfigured by altering (i) intramolecular base pairing, (ii) interaction of the mRNA with RNA-binding proteins, and (iii) interactions among the involved proteins through posttranslational modifications.
doi:10.1371/journal.pbio.0060092.g006

[29], further supports the idea of a mechanistic difference between translation termination at a PTC and termination at a "normal" TC.

Our finding that NMD suppression mediated through the poly(A) tail gradually declines with increasing distance between the TC and the poly(A) tail (Figure 3) is consistent with evidence suggesting that Upf1 and PABPC1 competes for interaction with release factor eRF3 bound to the ribosome at the TC [30]. Furthermore, our "distance model" provides a possible explanation for the reported distance effect of PABPC1 tethering to a β-globin NMD reporter transcript [31]. Consistent with our results on miniμ (Figures 3 and 5), NMD of this β-globin mRNA was suppressed more efficiently by tethering PABC1 45 nucleotides downstream of the PTC than tethering it 132 nucleotides downstream of the PTC [31]. The postulated requirement for correct translation termination to occur within a certain maximal physical distance from PABPC1 could also explain why PTCs near the start codon fail to trigger efficient NMD (Figure 1A and [32]), assuming that the start codon and the poly(A) tail are located in spatial proximity due to the interaction between eIF4G and PABPC1 [22].

Mammalian EJCs Have Evolved to Function as NMD Enhancers

Our data further show that in mammals, the EJC is not required for PTC recognition, as in *C. elegans* and *D. melanogaster* [20,33]. But unlike *C. elegans* and *D. melanogaster*, where the EJC does not appear to affect NMD at all, the EJC plays an important role as an enhancer of NMD in mammals. The comparison between NMD of miniμ ter440 and the miniμ ter440 C3/C4 (3'-most intron deleted) represents an example for such an EJC-mediated enhancement of NMD [17]. The simplest mechanistic explanation for the NMD-enhancing effect of EJCs is that they accelerate SMG1/Upf2/Upf3-dependent phosphorylation of Upf1 by locally concentrating Upf2 and Upf3 [34]. We demonstrated here that even in a situation of such EJC-enhanced NMD, positioning the poly(A) tail between the PTC and the EJC still strongly suppressed NMD (Figure 4), indicating that the translation termination-promoting signal in this position still efficiently competed with NMD-promoting events. Confirming our result and showing that PABPC1 is necessary for this NMD suppression, Ivanov and colleagues found that artificially inducing NMD by tethering the EJC factor Y14 into the 3' UTR of a β-globin reporter mRNA was suppressed by tethering PABPC1 between the TC and Y14 [35].

Noteworthy in the context of suppressing EJC-enhanced NMD, the mRNA decay kinetics of the reporter constructs in Figure 4B do not fit a simple exponential decay curve, but rather these transcripts seem to be degraded in an apparently bi-phasic mode. The initially fast decay rate might reflect the NMD enhancing effect of the EJC during the pioneer round of translation [25], whereas the second, somewhat slower decay rate would signify EJC-independent decay. We hypothesize that in mammals, where a large number of nonsense transcripts is produced by extensive alternative splicing [36], the EJC has evolved to function as an enhancer of NMD by locally concentrating Upf2 and Upf3b nearby terminating ribosomes and thereby tilting the balance of the two antagonizing signals toward NMD.

Implications for Disease-Associated Mutations

Collectively, these data have potentially important clinical implications. Our findings predict that mutations leading to extended 3′ UTRs, such as poly(A) site mutations or sequence insertions into the 3′ UTR, constitute a so far overlooked group of NMD substrates that may explain the molecular mechanism of certain genetic diseases. For such mutations, treatments with readthrough-promoting drugs like PTC124 [29] would not be suitable, because PTC124 does not stabilize the mRNA and would lead to the synthesis of C-terminally extended wild-type protein. In contrast, mRNA stabilization by a FB strategy as described here would augment wild-type protein levels and therefore represent a putative gene-specific therapeutic approach, provided the FB can be induced in trans.

Posttranscriptional Gene Regulation by Means of NMD

It has not escaped our notice that the new, unified NMD model we postulate immediately suggests a possible mechanism for posttranscriptional regulation of a wide variety of genes by NMD. It is well documented that 3′ UTRs, many of which comprise several thousand nucleotides in mammals, serve as binding sites for numerous factors that regulate mRNA translation or stability [26,27]. We postulate that by binding to their target transcript, many of these factors alter the tertiary structure of the 3′ UTR, thereby changing the local environment for translation termination (i.e., the physical distance between the TC and the poly(A) tail), which in turn will amend the transcript's half-life (Figure 6). 3′ UTR-binding factors can change the 3-D 3′ UTR configuration by masking mRNA sequences otherwise engaged in intramolecular base pairing, or by interacting with each other and thereby looping out mRNA sequence in-between. Protein-protein and protein-RNA interactions can be regulated through signal transduction pathways by posttranslational modification of the involved RNA-binding proteins. Furthermore, this NMD-dependent posttranscriptional gene regulation can also be modulated through transcript-specific RNA-binding proteins with intrinsic NMD-promoting or translation termination–promoting activity that binds into the proximity of the TC. For example, the RNA-binding protein Staufen has been reported to bind the 3′ UTR of a few specific mRNAs and to induce their rapid degradation by directly recruiting Upf1 [37,38]. Although it remains to be further investigated to which extent cells use this gene regulation pathway, the surprisingly large number of physiological transcripts detected by microarray analysis that raise in levels upon Upf1 knockdown indicates that it might be widespread [39–43]. A central prediction of this NMD-mediated gene regulation mechanism is that it depends on ongoing translation and that one would expect the set of transcripts affected by Upf1 depletion to vary in a tissue-specific manner, during development and differentiation, and by environmental cues in general. This might explain why the sets of transcripts affected by Upf1 depletion in the different microarray studies showed only limited overlap [39–43]. To test the postulated mode of gene regulation by spatial remodeling of the 3′ UTR more directly, development of techniques that allow in vivo measurements of the physical distance between two molecules and improved predictions of mRNA folding in the presence of RNA-binding proteins will be necessary.

Materials and Methods

Plasmids. The Ig-μ minigenes miniμWT, ter32, ter57, ter73, ter108, ter310, ter440, ter452, and ter459 are described elsewhere [44]. Nonsense mutations at amino acid positions 190 (CAG to TAG), 472 (TAC to TAG), and 517 (TAT to TAA) were generated by PCR-mediated site-directed mutagenesis using QuikChange XL. Site-Directed Mutagenesis Kit (Stratagene). The sequence of the entire ORF in each construct was confirmed by sequencing. For the miniμ constructs with extended 3′ UTRs, a unique ClaI site 90 bp downstream of the TC and a unique SpeI site 16 bp upstream of the poly(A) signal were generated in pβminiμ WT C3/C4 [17] by site-directed mutagenesis. The Ig-μ minigenes with extended 3′ UTR were generated by inserting 600-bp long PCR fragments of the β-lactamase (Amp) gene into the SpeI site (+600) or additionally, a second, similar PCR fragment into the ClaI site (+1,200). For the foldback constructs, a unique SpeI site was generated in pβminiμWT 16 bp upstream of the poly(A) signal by site-directed mutagenesis and its BamHI fragment was exchanged with the BamHI fragment of miniμ ter440 C4/H4 or miniμWT C4/H4 [17]. Insertion into the SpeI site of these plasmids of a ds-oligo with the sequence 5′-ctagCATTGGGTTTTGA-GATGAATTTCTTC-3′ gave miniμ ter440 C3/H4 FB and miniμ WT C3/H4 FB, and insertion of the ds-oligo in the opposite orientation gave miniμ ter440 C3/H4 NFB and miniμ WT C3/H4 NFB, respectively. The plasmids miniμ ter440 FB2, FB3, FB4, and FB5 were generated correspondingly by inserting the sequence 5′-ctagTGCCAGACA-TAGCTAATGTAATCTGAAAC-3′ (FB2), 5′-ctagCCTGGATGTTGT-CACGCAAGAC-3′ (FB3), 5′-ctagAGGAACACCTTCAGCACACCAC-3′ (FB4), and 5′-ctagCAAAACCAGTGACGTTTGAATGG-3′ (FB5) into the SpeI site. The plasmids miniμ ter310 FB, ter310 NFB, and WT NFB were generated analogously by inserting the sequence 5′-ctagTCGATTTCAGAGATGGTAAGTGTGC-3′ or the reverse complement into SpeI and miniμ ter310 and miniμ WT. The plasmid pβminiμter310 SL was generated by insertion of the sequence above into the BsrGI site of exon C2. The nucleotide substitutions to create the mutant FB constructs, numbered relative to the AUG start codon, were C964A, T965G, C968G, C969G, C974A, T975G, and G976C for miniμ ter310 FB and T1369G, T1370G, C1371A, T1375G, C1376A, C1381G, and C1382A for miniμ C3/H4 ter440 FB (sequence available upon request). pBS β-globin WT was described previously [45]. A unique MluI site two nucleotides downstream of the TC was generated by site-directed mutagenesis, and a 577-bp PCR fragment of the β-lactamase (Amp) gene with flanking MluI sites was inserted to generate pBS β-globin WT +577. For the half-life measurements, the entire reporter constructs from KpnI (in the 5′ UTR) to about 190 bp downstream of the polyA signal were inserted into KpnI-XhoI of pTRE-Tight (Clontech).

For the miniμ and TCRβ reporter gene constructs used in the tethering assay, a fragment encoding six MS2 binding sites from pcβwt β2-3MS2 [11] was PCR-amplified with flanking BsrGI sites and inserted either into the BsrGI site of exon C2 (giving the construct A) or into the BsrGI site of exon C4 (giving the construct B) of miniμ ter310. For TCR-β the six MS2 binding sites were introduced in the StuI (construct A) or BamHI (construct B) site of the TCR-β construct pβ434 [46]. The control plasmid containing the MS2 binding domain and a HA-tag was derived from pCMV-H2b-MS2-HA by replacing the H2b encoding KpnI–BamHI fragment with the annealed oligos 5′-CACCATG-3′ and 5′-GATCCATGGTGGTAC-3′ to introduce the Kozak consensus sequence. PABPC1 and PABPN1 were PCR-amplified from pETNHis-hPABPC1 (obtained from Elmar Wahle, University of Halle-Wittenberg, Germany) and pcDNA3-hPABPN1-HA [47] with flanking KpnI–BamHI sites and cloned into KpnI–BamHI of pCMV-H2b-MS2-HA to give pCMV-PABPC1-MS2-HA and pCMV-PABPN1-MS2-HA, respectively. The MS2-HA fragment was removed by digestion with BamHI and NotI and the HA-tag was then reintroduced by ds-oligos to generate pCMV-PABPC1-HA. The C-terminal part of the lacZ gene (1,910 bp) was PCR-amplified and inserted into KpnI-BamHI as described above. pSUPERpuro-hUpf1 and pSUPERpuro-scrambled constructs were described elsewhere [17,48]. pSUPERpuro-hUpf2 was obtained by transferring the EcoRI–HindIII fragment of pSUPER-hUpf2 [43] into pSUPERpuro. pSUPERpuro-hUpf3b-1 and pSUPERpuro-hUpf3b-2 were generated by insertion of double-stranded oligos encoding for short hairpin RNAs (shRNAs) into pSUPERpuro between the BglII and HindIII sites as described previously [49]. The two target sequences of hUpf3b were 5′-GGTGGTAATTCGAAGATTA-3′ (pSU-PERpuro-hUpf3b-1) and 5′-CGAGATCAGGAGCGCATAC-3′ (pSU-PERpuro-hUpf3b-2).

Cell culture and cycloheximide treatment. HeLa cells were grown in Dulbecco's modified Eagle's medium (DMEM, Invitrogen), supplemented with 10% heat-inactivated fetal calf serum (FCS), 100 U/mL penicillin, and 100 µg/mL streptomycin (Amimed). For the experiment in Figure S5, 44 h after transfection of the reporter plasmids, 100 µg/mL cycloheximide was added for 4 h before RNA isolation.

Half-life measurements. The Tet-Off Advanced Transactivator (tTA-Advanced) was stably integrated into the genome of HeLa cells according to the manufacturer's protocol (Tet-Off Advanced Inducible Gene Expression System, Clontech) and a cell clone with high tTA-Advanced expression was selected for further experiments. To determine decay kinetics of the miniµ reporter mRNAs, 2×10^5 per well of the tTA-Advanced expressing cells were seeded in 6-well plates. The next day, two wells for each time course were transfected with 100 ng pTRE-Tight miniµ reporter plasmid and 100 ng pBS β-globin WT plasmid (for normalization) per well, using 2 µl DreamFect (OZ Biosciences) according to the manufacturer's protocol. On the following day, the cells of these two wells were split into six wells. Time course was started 40 h after transfection by adding 1 µg/ml doxycycline (Sigma) to each well and harvesting the cells after 0, 1, 2, 4, 6, and 8 h.

Transient transfections and quantitative real-time RT-PCR. $2-3 \times 10^5$ HeLa cells were seeded in 6-well plates and transfected the next day with 100–150 ng reporter plasmid, 100–150 ng plasmid encoding a gene for normalization, and 2–4 µL DreamFect (OZ Biosciences) according to manufacturer's protocol. For normalization, pBS β-globin WT was used in all experiments except for Figures 5 and S2, where pCMVrGPx1-TGC [50] was used instead. For the tethering experiments in Figure 5, 300 ng MS2 fusion protein-encoding plasmid was cotransfected with 100 ng of reporter plasmid, and the cells were harvested 52 h after transfection. Total cellular RNA was isolated using "Absolutely RNA RT-PCR Miniprep Kit" (Stratagene) and 1 µg RNA was reverse transcribed in 50 µL Stratascript first strand buffer in the presence of 0.4 mM dNTPs, 300 ng random hexamers, 40 U RNasIn (Promega), and 50 U Stratascript reverse transcriptase (Stratagene) according to manufacturer's protocol. For Figures 1A, 2C, 4C, and S2. For Figures 1B–1D, 2B, 3, 4B, 5, S4, and S5, 1 µg RNA was reverse transcribed in 20 µL StrataScript 6.0 RT buffer in the presence of 1 mM dNTPs, 300 ng random hexamers, 40 U RNasIn (Promega), and 50 U StrataScript 6.0 reverse transcriptase (Stratagene) according to manufacturer's protocol. Real-time RT-PCR was performed as previously described [17].

RNAi. Knockdown of hUpf1, hUpf2, and hUpf3b was induced by transfection of pSUPERpuro plasmids targeting two different sequences in hUpf1 [48], one sequence in hUpf2 [43], or two different sequences in hUpf3b (see above), respectively. Starting 24 h after transfection, untransfected cells were eliminated by culturing the cells in the presence of 1.5 µg/mL puromycin for 48 h. Cells were then washed in PBS and incubated in puromycin-free medium for another 24 h. Total cellular RNA was isolated and whole cell lysates for Western blotting were prepared 96 h post transfection. The efficiency of the knockdown was assessed on the mRNA level by real-time RT-PCR (unpublished data) and on the protein level by Western blotting.

Northern blot analysis. Total cellular RNA (10 µg) was separated on a 1.2% agarose gel containing 1× MOPS and 1% formaldehyde. RNA was transferred to positively charged nylon membrane (Roche) in 0.5× MOPS by 1-h wet blotting in a genie blotter (Idea Scientific). Following UV crosslinking of the RNA to the nylon filter, pre-hybridization and hybridization of the blot in Figure 1B was carried out in 6× SSC, 5× Denhardt's reagent, and 0.5% SDS with 50 µg/mL denatured salmon sperm DNA and 100 µg/mL denatured calf thymus DNA at 60 °C. For hybridization, 100 ng µter310 and 20 ng β-globin DNA was labeled with α-^{32}P-dCTP using the Ready-To-Go DNA-Labeling Kit (Amersham). For Figure S3 was hybridized with an in vitro-transcribed, α-^{32}P-UTP-labeled antisense miniµ RNA probe in ULTRAHyb buffer (Ambion) at 68 °C. After overnight hybridization, membranes were washed twice with 2× SSC/0.2% SDS and twice with 0.2× SSC/0.1% SDS at 60 °C before exposure to a PhosphorImager screen.

Immunoblotting. Whole cell lysates corresponding to $0.37 \times 10^4 - 2 \times 10^5$ cells per lane were electrophoresed on a 10% SDS-PAGE. Proteins were transferred to Optitran BA-S 85 reinforced nitro-cellulose (Schleicher and Schuell) and probed with 1:2,500 diluted polyclonal rabbit anti-hUpf1, anti-hUpf2, or anti-hUpf3b antiserum [11], 1:1,000 diluted monoclonal mouse anti-lamin A/C (Santa Cruz Biotechnology) or anti-HA antibody (Roche), 1:400 diluted supernatant of the mouse hybridoma cell line Y12, which produces a monoclonal antibody against the human Sm B/B'proteins [51]. 1:2,500 diluted HRP-conjugated anti-rabbit IgG or HRP-conjugated anti-mouse IgG (Promega) was used as secondary antibody. ECL+ Plus

Western blotting detection system (Amersham) was used for detection and signals were visualized on a Luminescent Image Analyzer LAS-1000 (Fujifilm).

Supporting Information

Figure S1. Monitoring of the Knockdown Efficacy of Upf1, Upf2, and Upf3b at the mRNA Level

From the RNA samples of Figure 1D, relative mRNA levels of Upf1, Upf2, and Upf3b, normalized to endogenous GAPDH mRNA, were measured by RT-qPCR using the TaqMan assay Hs00161289__m1, Hs00210187__m1, Hs00224875__m1, and 432-6317E from Applied Biosystems. Average values of two qPCR runs are shown.

Found at doi:10.1371/journal.pbio.0060092.sg001 (214 KB PDF).

Figure S2. Extension of the 3′ UTR Converts the β-globin WT mRNA into a NMD Substrate

Relative β-globin mRNA levels from constructs with (+577) or without (control) a 3′ UTR extension, normalized to the mRNA levels of a cotransfected rGPx-1 gene, were determined in cells depleted (Upf1 k.d.) or not (scr k.d.) for Upf1. The effect of Upf1 knockdown on β-globin mRNA is shown in the histogram, and the efficacy of Upf1 depletion was monitored by Western blotting (lower panel). Average values and SD are from two independent experiments with three qPCR runs each. k.d., knockdown.

Found at doi:10.1371/journal.pbio.0060092.sg002 (362 KB PDF).

Figure S3. Detection of Miniµ mRNA by Northern Blot Analysis

(A) Miniµ C3H4 mRNA of FB and control NFB constructs from the RNA samples with Upf1 knockdown analyzed in Figure 2C.
(B) Miniµ mRNA of the constructs shown in Figure 4A. RNA samples of cycloheximide-treated cells (analyzed in Figure S5) were used.

Found at doi:10.1371/journal.pbio.0060092.sg003 (336 KB PDF).

Figure S4. Mutant FB Constructs Do Not Suppress NMD

Mutations in the sequence downstream of the PTC were introduced to abolish the folding back of the poly(A) tail of the miniµ C3H4 ter440 FB construct (A) and of the miniµ ter310 FB construct (D). The mutations are depicted in red. (B) and (E) Relative miniµ mRNA levels normalized to β-globin WT mRNA from a cotransfected plasmid were measured by RT-qPCR in cells depleted for Upf1 (Upf1 k.d.) or not (scr k.d.). Average mRNA levels and SD from one experiment with three RT-qPCR runs are shown and displayed as in the corresponding Figures 2C and 4C. (C) and (F) The efficacy of the Upf1 knockdown was assessed by Western blotting. SmB/B′ was detected as loading control.

Found at doi:10.1371/journal.pbio.0060092.sg004 (622 KB PDF).

Figure S5. Suppression of EJC-Enhanced NMD by Poly(A) Tail FB

Relative miniµ mRNA levels from the constructs indicated in Figure 4A, normalized and displayed as in Figure 4C, from cells treated (+CHX) or not (control) with 100 µg/mL cycloheximide for 4 h before RNA isolation. Average values and SD of five qPCR measurements from two independent experiments are shown.

Found at doi:10.1371/journal.pbio.0060092.sg005 (241 KB PDF).

Acknowledgments

We thank Jens Lykke-Andersen for communication of results prior to publication, for critical comments on the manuscript, and for antibodies and plasmids. We are also grateful to Andreas Kulozik, Lynne Maquat, Miles Wilkinson, Elmar Wahle, and Maria Carmo-Fonseca for plasmids.

Author contributions. ABE, LS, HM, and OM conceived and designed the experiments. ABE, LS, HM, RZO, and OM performed the experiments. ABE, LS, HM, and OM analyzed the data. OM, supported by ABE and LS, wrote the paper.

Funding. This work was supported by the Kanton Bern and by grants to OM from the Swiss National Science Foundation, the Novartis Foundation for Biomedical Research, and the Helmut Horten Foundation. OM is a fellow of the Max Cloëtta Foundation, and RZO is supported by a fellowship from CONACYT, México.

Competing interests. The authors have declared that no competing interests exist.

References

1. Isken O, Maquat LE (2007) Quality control of eukaryotic mRNA: safeguarding cells from abnormal mRNA function. Genes Dev 21: 1833-3856.
2. Chang YF, Imam JS, Wilkinson MF (2007) The nonsense-mediated decay RNA surveillance pathway. Annu Rev Biochem 76: 15.1-15.24.
3. Frischmeyer PA, Dietz HC (1999) Nonsense-mediated mRNA decay in health and disease. Mol Genet Hum 8: 1893-1900.
4. Holbrook JA, Neu-Yilik G, Hentze MW, Kulozik AE (2004) Nonsense-mediated decay approaches the clinic. Nat Genet 36: 801-808.
5. Rehwinkel J, Raes J, Izaurralde E (2006) Nonsense-mediated mRNA decay: target genes and functional diversification of effectors. Trends Biochem Sci 31: 639-646.
6. Culbertson MR, Leeds PF (2003) Looking at mRNA decay pathways through the window of molecular evolution. Curr Opin Genet Dev 13: 207-214.
7. Bhattacharya A, Czaplinski K, Trifillis P, He F, Jacobson A, et al. (2000) Characterization of the biochemical properties of the human Upf1 gene product that is involved in nonsense-mediated mRNA decay. RNA 6: 1226-1235.
8. Weng Y, Czaplinski K, Peltz SW (1996) Genetic and biochemical characterization of mutations in the ATPase and helicase regions of the Upf1 protein. Mol Cell Biol 16: 5477-5490.
9. Cheng Z, Muhlrad D, Lim MK, Parker R, Song H (2007) Structural and functional insights into the human Upf1 helicase core. EMBO J 26: 253-264.
10. He F, Brown AH, Jacobson A (1997) Upf1p, Nmd2p, and Upf3p are interacting components of the yeast nonsense-mediated mRNA decay pathway. Mol Cell Biol 17: 1580-1594.
11. Lykke-Andersen J, Shu MD, Steitz JA (2000) Human Upf proteins target an mRNA for nonsense-mediated decay when bound downstream of a termination codon. Cell 103: 1121-1131.
12. Mendell JT, Medghalchi SM, Lake RG, Noensie EN, Dietz HC (2000) Novel Upf2p orthologues suggest a functional link between translation initiation and nonsense surveillance complexes. Mol Cell Biol 20: 8944-8957.
13. Serin G, Gersappe A, Black JD, Aronoff R, Maquat LE (2001) Identification and characterization of human orthologues to Saccharomyces cerevisiae Upf2 protein and Upf3 protein (Caenorhabditis elegans SMG-4). Mol Cell Biol 21: 209-223.
14. Yamashita A, Kashima I, Ohno S (2005) The role of SMG-1 in nonsense-mediated mRNA decay. Biochim Biophys Acta 1754: 305-315
15. Lejeune F, Maquat LE (2005) Mechanistic links between nonsense-mediated mRNA decay and pre-mRNA splicing in mammalian cells. Curr Opin Cell Biol 17: 309-315.
16. Behm-Ansmant I, Izaurralde E (2006) Quality control of gene expression: a stepwise assembly pathway for the surveillance complex that triggers nonsense-mediated mRNA decay. Genes Dev 20: 391-398.
17. Buhler M, Steiner S, Mohn F, Paillusson A, Muhlemann O (2006) EJC-independent degradation of nonsense immunoglobulin-mu mRNA depends on 3' UTR length. Nat Struct Mol Biol 13: 462-464.
18. Amrani N, Ganesan R, Kervestin S, Mangus DA, Ghosh S, et al. (2004) A faux 3'-UTR promotes aberrant termination and triggers nonsense-mediated mRNA decay. Nature 432: 112-118.
19. Behm-Ansmant I, Gatfield D, Rehwinkel J, Hilgers V, Izaurralde EA (2007) Conserved role for cytoplasmic poly(A)-binding protein 1 (PABPC1) in nonsense-mediated mRNA decay. EMBO J 26: 1591-1601.
20. Longman D, Plasterk RH, Johnstone IL, Caceres JF (2007) Mechanistic insights and identification of two novel factors in the C. elegans NMD pathway. Genes Dev 21: 1075-1085.
21. Amrani N, Sachs MS, Jacobson A (2006) Early nonsense: mRNA decay solves a translational problem. Nat Rev Mol Cell Biol 7: 415-425.
22. Wells SE, Hillner PE, Vale RD, Sachs AB (1998) Circularization of mRNA by eukaryotic translation initiation factors. Mol Cell 2: 135-140.
23. Lejeune F, Ranganathan AC, Maquat LE (2004) eIF4G is required for the pioneer round of translation in mammalian cells. Nat Struct Mol Biol 11: 992-1000.
24. Fortes P, Inada T, Preiss T, Hentze MW, Mattaj IW, et al. (2000) The yeast nuclear cap binding complex can interact with translation factor eIF4G and mediate translation initiation. Mol Cell 6: 191-196.
25. Ishigaki Y, Li X, Serin G, Maquat LE (2001) Evidence for a pioneer round of mRNA translation: mRNAs subject to nonsense-mediated decay in mammalian cells are bound by CBP80 and CBP20. Cell 106: 607-617.
26. Wickens M, Anderson P, Jackson RJ (1997) Life and death in the cytoplasm: messages from the 3' end. Curr Opin Genet Dev 7: 220-232.
27. Moore MJ (2005) From birth to death: the complex lives of eukaryotic mRNAs. Science 309: 1514-1518.
28. Hilleren P, Parker R (1999) mRNA surveillance in eukaryotes: kinetic proofreading of proper translation termination as assessed by mRNP domain organization? RNA 5: 711-719.
29. Welch EM, Barton ER, Zhuo J, Tomizawa Y, Friesen WJ, et al. (2007) PTC124 targets genetic disorders caused by nonsense mutations. Nature 447: 87-91.
30. Singh G, Rebbapragada I, Lykke-Andersen JA (2008) Competition between stimulators and antagonists of Upf complex recruitment governs human nonsense-mediated mRNA decay. PLoS Biol 6: e111. doi:10.1371/journal.pbio.0060111
31. Silva AL, Ribeiro P, Inacio A, Liebhaber SA, Romao L (2008) Proximity of the poly(A)-binding protein to a premature termination codon inhibits mammalian nonsense-mediated mRNA decay. RNA 14: 563-576.
32. Silva AL, Pereira FJ, Morgado A, Kong J, Martins R, et al. (2006) The canonical UPF1-dependent nonsense-mediated mRNA decay is inhibited in transcripts carrying a short open reading frame independent of sequence context. RNA 12: 2160-2170.
33. Gatfield D, Unterholzner L, Ciccarelli FD, Bork P, Izaurralde E (2003) Nonsense-mediated mRNA decay in Drosophila: at the intersection of the yeast and mammalian pathways. EMBO J 22: 3960-3970.
34. Kashima I, Yamashita A, Izumi N, Kataoka N, Morishita R, et al. (2006) Binding of a novel SMG-1-Upf1-eRF1-eRF3 complex (SURF) to the exon junction complex triggers Upf1 phosphorylation and nonsense-mediated mRNA decay. Genes Dev 20: 355-367.
35. Ivanov PV, Gehring NH, Kunz JB, Hentze MW, Kulozik AE (2008) Interactions between UPF1, eRFs, PABP, and the exon junction complex suggest an integrated model for mammalian NMD pathways. EMBO J 27: 736-747.
36. Lewis BP, Green RE, Brenner SE (2003) Evidence for the widespread coupling of alternative splicing and nonsense-mediated mRNA decay in humans. Proc Natl Acad Sci U S A 100: 189-192.
37. Kim YK, Furic L, Desgroseillers L, Maquat LE (2005) Mammalian Staufen1 recruits Upf1 to specific mRNA 3' UTRs so as to elicit mRNA decay. Cell 120: 195-208.
38. Kim YK, Furic L, Parisien M, Major F, DesGroseillers L, et al. (2007) Staufen1 regulates diverse classes of mammalian transcripts. EMBO J 26: 2670-2681.
39. He F, Li X, Spatrick P, Casillo R, Dong S, et al. (2003) Genome-wide analysis of mRNAs regulated by the nonsense-mediated and 5' to 3' mRNA decay pathways in yeast. Mol Cell 12: 1439-1452.
40. Lelivelt MJ, Culbertson MR (1999) Yeast Upf proteins required for RNA surveillance affect global expression of the yeast transcriptome. Mol Cell Biol 19: 6710-6719.
41. Mendell JT, Sharifi NA, Meyers JL, Martinez-Murillo F, Dietz HC (2004) Nonsense surveillance regulates expression of diverse classes of mammalian transcripts and mutes genomic noise. Nat Genet 36: 1073-1078.
42. Rehwinkel J, Letunic I, Raes J, Bork P, Izaurralde E (2005) Nonsense-mediated mRNA decay factors act in concert to regulate common mRNA targets. RNA 11: 1530-1544.
43. Wittmann J, Hol EM, Jack HM (2006) hUPF2 silencing identifies physiologic substrates of mammalian nonsense-mediated mRNA decay. Mol Cell Biol 26: 1272-1287.
44. Buhler M, Paillusson A, Muhlemann O (2004) Efficient downregulation of immunoglobulin mu mRNA with premature translation-termination codons requires the 5'-half of the VDJ exon. Nucleic Acids Res 32: 3304-3315.
45. Thermann R, Neu-Yilik G, Deters A, Frede U, Wehr K, et al. (1998) Binary specification of nonsense codons by splicing and cytoplasmic translation. EMBO J 17: 3484-3494.
46. Mohn F, Buhler M, Muhlemann O (2005) Nonsense-associated alternative splicing of T cell receptor beta genes: no evidence for frame dependence. RNA 11: 147-156.
47. Tavanez JP, Calado P, Braga J, Lafarga M, Carmo-Fonseca M (2005) In vivo aggregation properties of the nuclear poly(A)-binding protein PABPN1. RNA 11: 752-762.
48. Paillusson A, Hirschi N, Vallan C, Azzalin CM, Muhlemann O (2005) A GFP-based reporter system to monitor nonsense-mediated mRNA decay. Nucleic Acids Res 33: e54.
49. Brummelkamp TR, Bernards R, Agami RA (2002) System for stable expression of short interfering RNAs in mammalian cells. Science 296: 550-553.
50. Moriarty PM, Reddy CC, Maquat LE (1997) The presence of an intron within the rat gene for selenium-dependent glutathione peroxidase 1 is not required to protect nuclear RNA from UGA-mediated decay. RNA 3: 1369-1373.
51. Lerner EA, Lerner MR, Janeway CA Jr, Steitz JA (1981) Monoclonal antibodies to nucleic acid-containing cellular constituents: probes for molecular biology and autoimmune disease. Proc Natl Acad Sci U S A 78: 2737-2741.

6.3.4. Paper IV

Stalder L, Mühlemann O. The meaning of nonsense. Trends Cell Biol. 2008 Jul;18(7):315-21.

Appendix: Paper IV

Opinion

The meaning of nonsense

Lukas Stalder and Oliver Mühlemann

Institute of Cell Biology, University of Berne, Baltzerstraße 4, 3012 Berne, Switzerland

To ensure the accuracy of gene expression, eukaryotes have evolved several surveillance mechanisms. One of the best-studied quality control mechanisms is nonsense-mediated mRNA decay (NMD), which recognizes and degrades transcripts harboring a premature translation-termination codon (PTC), thereby preventing the production of faulty proteins. NMD regulates ~10% of human mRNAs, and its physiological importance is manifested by the fact that ~30% of disease-associated mutations generate PTCs. Although different mechanisms of PTC recognition have been proposed for different species, recent studies in Saccharomyces cerevisiae, Drosophila melanogaster, Caenorhabditis elegans, plants and mammals suggest a conserved model. Here, we summarize the latest results and discuss an emerging model for NMD and its implications for the regulation of gene expression.

Introduction

Eukaryotic gene expression involves an intricate chain of complex biochemical reactions, starting with the synthesis of mRNA, followed by the production of encoded proteins, and ending with the degradation of both the mRNA and the protein. Tight control of and high accuracy within these processes are absolutely required to prevent inappropriate gene expression and to ensure cell survival, and cells have therefore evolved mechanisms to control many steps along the chain [1]. One of the best-studied quality control mechanisms is nonsense-mediated mRNA decay (NMD). NMD was initially described as a mechanism for recognizing and degrading faulty transcripts harbouring a premature translation-termination codon (PTC); such nonsense transcripts would otherwise result in the production of C-terminally truncated proteins with potentially dominant–negative effects. PTCs can arise either from mutations at the DNA level (e.g. nonsense mutations, frame-shifting deletions and insertions) or from altered splicing signals that induce production of alternatively spliced mRNA isoforms with truncated reading frames [2]. It has been estimated that among the 60–70% of pre-mRNAs that undergo alternative splicing, 45% generate at least one splice form predicted to be an NMD substrate [3]. Over the last five years, it has become clear that NMD not only degrades faulty transcripts but also regulates the steady-state level of many physiological mRNAs involved in a variety of different cellular processes, such as DNA repair, the cell cycle, and metabolism [4–6]. Genome-wide screens in budding yeast, Drosophila and human cells have revealed that NMD regulates expression of ~3–10% of the transcriptome [4–6]. Furthermore, it has been estimated that ~30% of the known disease-associated mutations in humans generate a PTC-containing (PTC+) mRNA, and in many of these cases NMD influences the severity of the clinical manifestations caused by the mutation [7].

Despite intense investigation over the past two decades, the molecular mechanisms of NMD are still not fully understood. Indeed, different models to explain PTC recognition have emerged from studies in different organisms. However, recent data from Saccharomyces cerevisiae, Drosophila melanogaster, Caenorhabditis elegans, plants and mammals suggest that the basic mechanism for PTC recognition is much more conserved than previously thought [8–16]. On the basis of these new data, we discuss a 'unified NMD model', implicating a novel post-transcriptional mode of gene regulation. Furthermore, we highlight the implications of this NMD model for the clinical manifestation of genetic diseases.

Trans-acting factors involved in NMD

The NMD core factors Upf1p, Upf2p and Upf3p (for "up-frameshift 1-3") were initially identified in genetic screens in yeast, and SMG1-7 (for "suppressor with morphological effect on genitalia") were found to be NMD effectors in C. elegans (for review, see [17]). Sequence alignments revealed that SMG2 is homologous to Upf1p, SMG3 to Upf2p, and SMG4 to Upf3p, and that SMG1-7 are present in all higher eukaryotes analyzed to date (Figure 1), with the exception of D. melanogaster, which contains no clear homolog for SMG7 [17]. Additional NMD factors are very likely to exist, although they remain to be discovered. Recently, smg1 (also known as hNAG) and smg2 (also known as hDHX34) have been identified as NMD factors in C. elegans and humans [11].

Given that they are conserved from yeast to humans, UPF1, UPF2 and UPF3 are believed to function at the heart of NMD. UPF1 is the most highly conserved NMD factor, and elucidating its structure and function will provide the key to understanding the mechanism of NMD. UPF1 interacts with the eukaryotic release factors eRF1 and eRF3 (see Glossary) [8,18–20], binds to UPF2 through its cysteine and histidine rich (CH-rich) region near the N-terminus [21], and it interacts with SMG1, SMG5, SMG6 and SMG7 [19,22–27]. Furthermore,

---

Glossary

3′ UTR: Untranslated region downstream of the termination codon.
eRF1 and eRF3: Eukaryotic release factors 1 and 3, which function in translation termination.
P bodies: Processing bodies, also called GW182-containing bodies, DCP1 foci, or XRN1 foci.
PABPC1: the major cytoplasmic form of poly(A)-binding protein in mammals.

Corresponding author: Mühlemann, O. (oliver.muehlemann@izb.unibe.ch).

Opinion

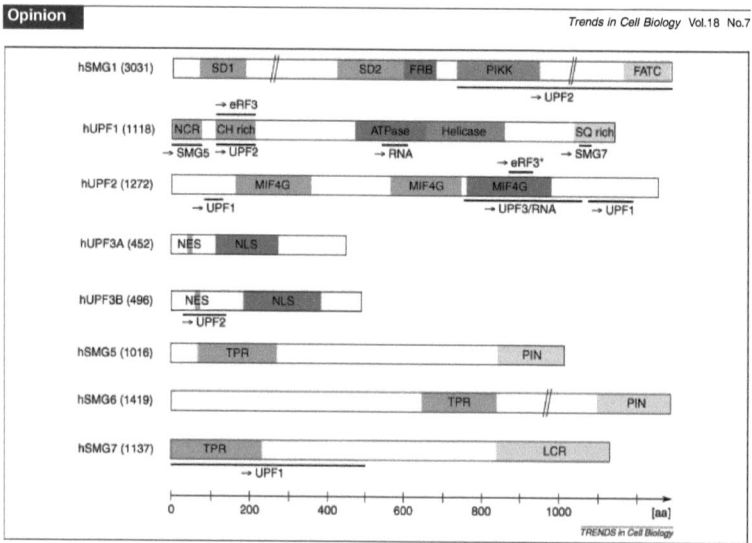

Figure 1. Human NMD factors. The known human NMD factors are depicted schematically. The proteins are drawn to scale with respect to the number of amino acids that they have, indicated by the scale bar at the bottom. Direct interactions with other proteins or RNA are shown as black bars. Reported interactions for which the interaction domains are not yet mapped are not indicated (e.g. UPF1–SMG1, SMG5–SMG7, UPF1–SMG6 and UPF2–SMG1 [19,25,28]). Two SMG1 homology domains, SD1 and SD2, and an FKBP12-rapamycin binding site (FRB) have been mapped in hSMG1. hSMG1 also contains a FRAP–ATM–TRRAP C-terminal domain (FATC), which is found in the majority of PI3K-related kinases (PIKKs) [19,24,58]. hUPF1 contains an N-terminal conserved region (NCR), a CH-rich domain, a core domain comprising the ATPase and helicase, and an SQ-rich C-terminal domain [8,19,21,26,31,32]. hUPF2 contains three mammalian eIF4G-like (MIF4G) domains [19,20,32,59]. Regions containing nuclear export signals (NESs) and nuclear localization signals (NLSs) were found in hUPF3A and hUPF3B [20,23,59]. hSMG5, hSMG6 and hSMG7 contain two tetratricopeptide repeats (TPRs), which were shown to adopt a 14-3-3-like fold in hSMG7 [27]. The sequence similarity between the TPRs of hSMG5 and hSMG6 suggests that they also probably fold in a 14-3-3-like manner and bind to phosphorylated UPF1. hSMG5 and hSMG6 contain a C-terminal PIN domain (for PilT N terminus), which is present in proteins that have ribonuclease activity; a C-terminal low complexity region (LCR) was mapped in hSMG7, and this region is sufficient for the localization of the protein to P bodies and for its subsequent degradation, when tethered to a reporter mRNA [25–27,37,49,60]. The asterisk indicates regions that have only been mapped in S. cerevisiae.

UPF1 undergoes a cycle of phosphorylation and dephosphorylation, which is essential for NMD in metazoans and is regulated by the other NMD factors. SMG1 phosphorylates UPF1 at numerous serine residues in the C terminus [19,24]. This UPF1 phosphorylation was reported to depend on the presence of both UPF2 and UPF3b in humans [19], although recent studies provided evidence for UPF2- and UPF3b-independent NMD [8,28,29]. Phosphorylated UPF1 can be bound and dephosphorylated by SMG5, SMG6 and SMG7, all of which harbor two consensus tetratricopeptide repeat sequence motifs (TPRs) and recruit the phosphatase PP2A to UPF1 [22,25–27]. UPF1 exhibits RNA binding, RNA-dependent ATP hydrolysis, and 5′-to-3′ ATP-dependent RNA helicase activities, and inhibition of any of these activities suppresses NMD [19,30,31]. It has been shown *in vitro* that UPF1 dissociates from RNA after addition of ATP [30]. Interaction with eRF1 and eRF3 strongly reduces the ATPase and RNA binding activities of UPF1 [18]. Vice versa, RNA binding stimulates the ATPase activity and leads to a dissociation of UPF1 from eRF1 and eRF3. Furthermore, it was recently demonstrated that UPF2 and UPF3b cooperatively stimulate the ATPase and RNA helicase activity of UPF1 *in vitro* [32].

Proper versus aberrant translation termination

A central question to understand NMD is how a PTC can be distinguished from a natural termination codon (TC). Despite the conservation of the core NMD factors, remarkably different models have been proposed for different organisms [2,33,34]. In mammalian cells, the presence of an exon junction complex (EJC) downstream of the termination codon was found to be a crucial determinant for defining the termination codon as being premature (for review, see [2,34]). However, this model has been challenged by reports that NMD occurs even in the absence of a downstream EJC in mammals [15,16,35,36], and that NMD occurs independently of splicing in D. melanogaster and C. elegans, suggesting that the EJC is not involved in NMD in worms and flies [11,37].

An alternative model for PTC recogniton invokes a '*faux* 3′ untranslated region' (UTR; see Glossary). This model postulates that proper (or efficient) translation

Opinion

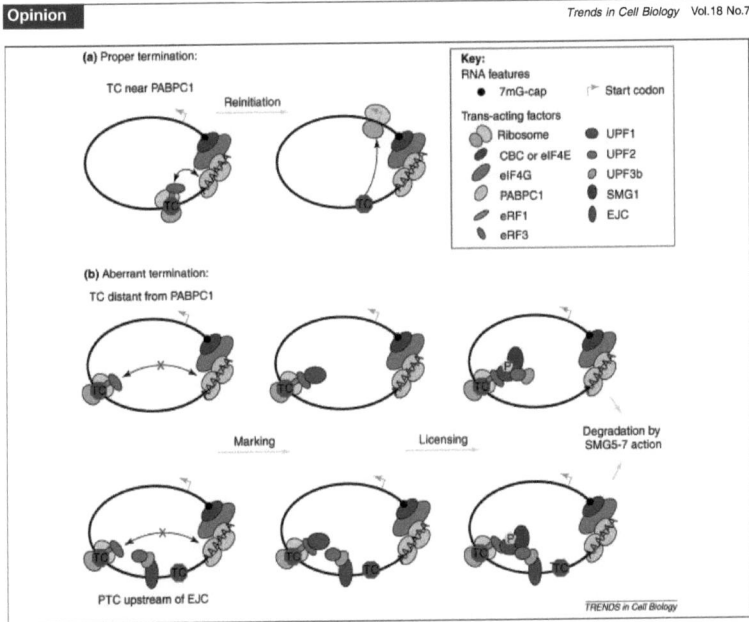

Figure 2. Model for PTC recognition. (a) The mRNP is thought to form a closed-loop structure through the interaction of cap-bound eIF4E or CBC (for "cap-binding complex") with eIF4G, which in turn interacts with the poly(A) tail-bound PABPC1. When the ribosome terminates at a TC in the vicinity of the poly(A) tail, a PABPC1-mediated signal promotes proper termination of translation, resulting in efficient reinitiation of the ribosome at the 5' end of the mRNA, and the production of a stable mRNP. (b) If the ribosome terminates at a TC that is too far away from the poly(A) tail for it to receive the PABPC1-mediated translation-termination-promoting signal, UPF1 binds to the stalled ribosome instead, thereby marking this TC as premature. Subsequently, UPF2 and UPF3b interact with UPF1, promoting SMG1-mediated phosphorylation of UPF1. This licensing step commits the mRNA to rapid degradation by as yet unknown pathways that involve the binding of SMG5-7 to the phosphorylated UPF1 (upper part). An EJC downstream of a TC functions as an NMD enhancer by shortening the time window between UPF1 binding and its phosphorylation by locally concentrating UPF2 and UPF3b (lower part).

termination requires a termination-promoting signal (Figure 2a), and that the absence of this signal typifies aberrant translation-termination at a PTC, which in turn leads to degradation of the mRNA (Figure 2b) [10,33]. Consistently, evidence for kinetic and mechanistic differences between aberrant and normal translation termination events has been obtained in *S. cerevisiae* [10]. Interestingly, tethering of poly(A) binding protein (Pab1p) into the proximity of a PTC suppressed NMD, suggesting that Pab1p or some factor bound to Pab1p might transmit the signal required for proper termination of translation [10]. Although the exact molecular events of translation termination remain to be elucidated, it seems that proper termination requires a certain mRNP structure and specific factors to promote efficient polypeptide release, ribosome disassembly, and recycling of the ribosome subunits to the 5' end of the mRNA [10,33]. We speculate that the closed-loop structure of an mRNP, adopted by juxtaposing the 5' and 3' ends through the eIF4E–eIF4G–Pab1p interaction, might represent such a structural environment for proper termination of translation [38]. The yeast data suggest that, if termination occurs too far away from this mRNP environment, disassembly of the ribosome is slow [10] and, as a result, Upf1p is recruited to the terminating ribosome through interaction with Sup35p and Sup45p, the yeast homologues of eRF3 and eRF1, respectively [18,33,39].

An evolutionarily conserved model for PTC recognition

Consistent with the *faux* 3' UTR model, deletions that eliminate most of the sequence downstream of a PTC abolish NMD [36]. Furthermore, mRNAs with a long 3' UTR were identified as being NMD substrates in *S. cerevisiae*, *D. melanogaster*, *C. elegans*, *Arabidopsis thaliana* and humans [9,11–13,15,16,36,40]. Recent studies in *D.*

Opinion

melanogaster and humans demonstrate that tethering poly(A) binding protein (PABP) downstream of, but close to, the PTC suppresses NMD. This finding, also previously observed in yeast, further corroborates the idea of a common, evolutionarily conserved mechanism for PTC recognition [8–10,14–16]. Based on these recent reports from different organisms, we propose here a 'unified' model for PTC recognition. This model essentially extends the *faux* 3′ UTR model to all species and proposes that, during mammalian NMD, downstream EJCs act as an evolutionary adaptation to efficiently recognize nonsense mRNAs produced by extensive alternative splicing. Our laboratory recently showed that the physical distance, rather than the number of nucleotides, between a TC and the poly(A) tail is a crucial determinant in defining a TC as premature [15]. Specifically, NMD of PTC-containing immunoglobulin-μ reporter transcripts expressed in human cells was suppressed by bringing the poly(A) tail into the vicinity of the PTC by means of a secondary structure. Furthermore, it was shown that UPF1 and PABPC1 (see Glossary), the major cytoplasmic PABP, compete for the interaction with eRF3 in human cells *in vitro* [16]. In our study, we found that the extent of NMD suppression in the fold-back constructs gradually declines with increasing distance between the poly(A) tail and the TC. This is consistent with the notion that the balance of the competition between PABPC1 and UPF1 for interaction with eRF3 enables the cell to distinguish between correct and aberrant translation termination [15]. Thus, if a ribosome stalls at a TC that is too far away from the termination-promoting environment (i.e. distant from PABPC1), resulting in slow termination kinetics, the balance between the two antagonizing signals is tilted toward UPF1 binding ('marking' within Figure 2b). Notably, binding of UPF1 to the stalled ribosome in this context is EJC-independent. Based on data from *S. cerevisiae, C. elegans, D. melanogaster, A. thaliana* and humans, we propose that this marking of the aberrant mRNP by UPF1 is conserved among eukaryotes and represents the fundamental step in PTC recognition.

Different second signals have evolved

Although UPF1 has been found to preferentially associate with PTC-containing mRNA in *C. elegans*, some association with PTC-free mRNA was also observed [41]. Furthermore, in *S. cerevisiae*, Upf1p can target normal mRNAs to P bodies (see Glossary; for review, see [42]) without promoting the degradation of these mRNAs [43]. This suggests that simple binding of UPF1 to a terminating ribosome is not sufficient to elicit degradation, but rather a 'second signal' is required. The second signal might be the binding of UPF2 and UPF3 to UPF1, an event that is important for the SMG1-mediated phosphorylation of UPF1 in higher eukaryotes [19,44], and which stimulates the RNA helicase and ATPase activities of UPF1 [19,32]. As a consequence of ATP hydrolysis, binding of UPF1 to the RNA might be facilitated. Supporting this view, a UPF1 mutant that cannot interact with UPF2 was shown to accumulate in its unphosphorylated form in a complex with SMG1, eRF1 and eRF3 [19]. Furthermore, ATPase-defective UPF1 mutants show enriched co-immunoprecipitation with UPF2 and UPF3b

in mammalian cells [19]. In summary, we postulate that the phosphorylation of UPF1 and the stimulation of its ATP hydrolysis and helicase activity represent a 'point of no return' in the NMD pathway, which we define this as the 'licensing step' (Figure 2b). In contrast to the situation in mammals, ATP hydrolysis appears to be required for the recruitment of Upf2p and Upf3p after Upf1p-dependent targeting of the mRNAs to P bodies in yeast [43]. In conjunction with the absence of SMG1, SMG5, SMG6 and SMG7 homologs in *S. cerevisiae*, we suggest that the NMD pathway for *S. cerevisiae* diverges from the pathways of metazoans at this licensing step.

Degradation of UPF1-bound mRNA

After the licensing step in metazoans, SMG5, SMG6 and SMG7 bind to the phosphorylated UPF1 through their 14-3-3-like domain, leading to the degradation of the mRNA (Figure 2b) [25–27]. However, the exact molecular relationship between UPF1 and SMG5–7 is still unclear.

In yeast and mammals, nonsense transcripts appear to be degraded by exosome-mediated 3′-5′ decay and by decapping followed by XRN1-mediated (for "5′-3′ exonuclease 1") 5′-3′ decay [45,46]. Several interactions between NMD and mRNA decay factors have been mapped consistently. For example, UPF1 interacts with the decapping enzyme in *S. cerevisiae* and human cells [47]. In *D. melanogaster*, however, the degradation of PTC-containing transcripts is initiated by endonucleolytic cleavage, and the resulting cleavage-fragments are subsequently degraded by 5′-3′ and 3′-5′ exonucleases [48].

The cellular localization of the degradation is not yet clear. Recent studies have shown that P bodies are sites of NMD in *S. cerevisiae* [43]. P bodies are implicated in cellular degradation in mammalian cells too, because they contain decapping and degradation enzymes, NMD factors, and effectors of the RNA interference (RNAi) silencing pathway (for review, see [42]). SMG7 might provide the molecular link between NMD and the degradation machinery in mammalian cells. When overexpressed, SMG7 accumulates in P bodies, which also leads to accumulation of SMG5 and UPF1 there. Furthermore, when SMG7 is tethered to a reporter transcript, it is able to elicit mRNA decay independent of a PTC [49].

EJC has evolved as an enhancer of NMD in mammals

Contrary to the popular model for mammalian NMD, several studies have demonstrated that PTCs can trigger NMD in the absence of an EJC further downstream on the mRNA [15,16,34,36,50,51]. However, it is apparent that the extent of mRNA downregulation in these examples of EJC-independent NMD is lower than in corresponding examples of NMD of transcripts with EJCs in the 3′ UTR. Consistent with the idea that EJCs have an important role in NMD, knockdown of EJC core-factors in mammalian cells reduced the downregulation of many NMD reporter mRNAs [36,52–54]. In the light of this, we propose that, in mammals, the EJC has evolved as a specialized second signal to enhance mammalian NMD. Our unified NMD model provides a mechanistic explanation for the NMD-enhancing function of EJCs located downstream of a TC. As part of such a 3′ UTR-bound EJC, UPF2 and UPF3 are ideally positioned for immediate interaction with ribosome-bound UPF1 and SMG1. As a consequence, the time window between the binding of UPF1 to the terminating ribosome (the marking step) and its SMG1-mediated phosphorylation (the licensing step) would be shortened, and thus the competition between PABP and UPF1 for binding to the stalled ribosome would tilt toward NMD (Figure 2b). We hypothesize that, in mammals, under the evolutionary pressure to efficiently recognize and eliminate the large number of nonsense mRNAs produced by extensive alternative pre-mRNA splicing, the EJC as a spatial mark of previous splicing events has been incorporated into the mechanism of PTC recognition as an enhancer. Consistent with this view, proteins homologous to mammalian EJC core components are not involved in NMD in *D. melanogaster* and *C. elegans* [11,37], in which only a minor fraction of pre-mRNA is alternatively spliced. Notably, downstream sequence elements (DSEs) identified in *S. cerevisiae* might have NMD-enhancing functions similar to that of the EJC, by providing a binding platform for NMD enhancing factors [55].

NMD as a novel mode of translation-dependent post-transcriptional gene regulation

3′ UTRs, which in mammals can comprise several thousand nucleotides, contain binding sites for numerous factors that are known to regulate mRNA translation and stability [1]. As mentioned previously, we recently found that, by changing the spatial configuration of the 3′ UTR of a transcript by introducing intramolecular base pairing, the half-life of the transcript can be changed in an NMD-dependent manner. This suggests that NMD has a novel role as a post-transcriptional mechanism for gene regulation (Figure 3) [15]. We predict that many RNA-binding factors alter the tertiary structure of the target transcript. Such structural rearrangements can, for example, change the physical distance between the TC and the poly(A) tail, and therewith change the local environment for translation termination; this in turn would affect the half-life of the mRNA and, as a consequence, have an effect at the protein level. RNA-binding factors can be either proteins or RNAs, and they can alter the 3-dimensional configuration of the 3′ UTR by masking mRNA sequences otherwise engaged in intramolecular base pairing, or by interacting with each other and thereby looping out mRNA sequences in-between. Importantly, such protein–protein and protein–RNA interactions can be regulated by environmental cues through signal transduction pathways that modify the involved RNA-binding proteins, for example by phosphorylation. In addition, transcript-specific RNA-binding proteins with intrinsic NMD-promoting or translation termination-promoting activities could directly modulate mRNA stability by binding near to the TC. To our knowledge, no physiological examples regulated by spatial remodeling of the 3′ UTR have been reported to date; however, there are likely to be many, because microarray analysis has revealed a surprisingly large number of physiological transcripts that rise in levels upon UPF1 knockdown [4–6,56]. Our NMD-mediated gene regulation model predicts that ongoing translation is required and that the population of transcripts affected by UPF1 depletion varies in a tissue-specific manner, during development and differentiation, and by environmental cues in general. Although experimental differences cannot be excluded, this might at least in part explain why the sets of transcripts affected by UPF1 depletion in the different microarray studies overlapped only very little [4–6,56,57]. Clearly, future work is needed to test the biological relevance of this model.

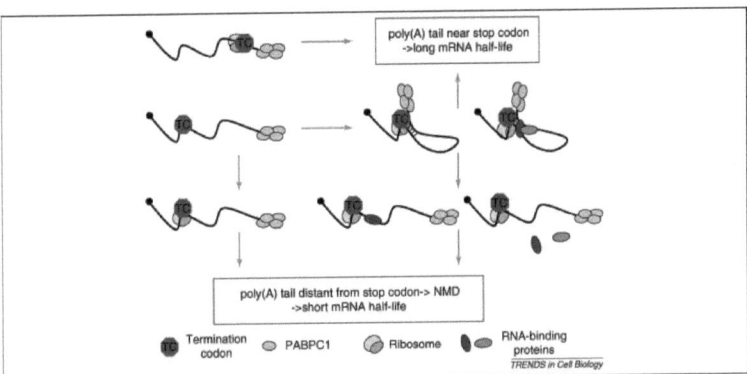

Figure 3. Post-transcriptional gene regulation by spatial rearrangement of the 3′ UTR. Many mammalian 3′ UTRs comprise thousands of nucleotides. They are most probably highly structured, and they provide binding sites for RNA binding proteins. We postulate that, by modulating the 3-dimensional configuration of the 3′ UTR and thereby the local environment for translation termination, the stability of mRNAs can be regulated in a translation-dependent manner. Such structural rearrangements can be mediated, for example, by RNA-binding factors that mask mRNA sequences that are otherwise involved in intramolecular base pairing, or by interacting with each other (Figure adapted from [15]).

Opinion

Genetic diseases: implications of the unified NMD model

It has been estimated that ~30% of the known disease-associated mutations generate a PTC, which suggests that NMD has a widespread impact on the phenotype of numerous genetic diseases [7]. NMD is beneficial if it prevents the production of C-terminally truncated proteins that would have had dominant–negative effects. By contrast, NMD is detrimental if it prevents the production of truncated proteins that still have residual function, as has been described for frequent mutations causing cystic fibrosis [7]. In addition to the existence of these mutations in the open reading frame of a gene, the unified NMD model predicts the existence of more populations of NMD targets than were previously appreciated. On the one hand, many PTCs in the last exon have the potential to elicit NMD, especially when the last exon is long. On the other hand, various mutations in the 3′ UTR have the potential to alter the spatial relationship between the TC and the poly(A) tail: insertions into the 3′ UTR; mutations that destroy poly(A) sites or create cryptic ones; and modification of binding sites for RNA binding proteins. Any of these types of mutations could turn an mRNA into an NMD target. Collectively, this suggests that the total number of genetic diseases influenced by NMD has been underestimated.

Conclusions and open questions

Although NMD has been investigated intensively for at least 20 years, the molecular mechanisms are not yet fully understood. Here, we summarize some recent publications showing that proper translation termination requires an interaction of the terminating ribosome with poly(A) binding protein, that the physical distance between the TC and the poly(A) tail is a crucial determinant for recognition of PTCs, and that PAPBC1 and UPF1 compete for the interaction with eRF3. Accordingly, when a ribosome stalls at a PTC, it lacks the PABP-mediated termination-promoting signals, and therefore UPF1 can out-compete PABP for the interaction with eRF3. This unified NMD model is consistent with recent findings in the other eukaryotes and suggests a conservation of the basic mechanism of PTC recognition. According to this new view, spatial remodeling of the 3′ UTR provides a novel mechanism whereby cells can regulate gene expression post-transcriptionally and in a translation-dependent manner. Although physiological examples regulated by this mechanism have not been reported to date, it is consistent with the finding that NMD not only rids the cell of faulty transcripts but also is involved in the regulation of 'normal' transcripts. The mechanisms proposed in this review could explain why NMD regulates such a large number of transcripts, many of which would not be predicted to be NMD substrates according to the currently prevailing model. Furthermore, it provides a possible explanation for the poor overlap between the sets of NMD-regulated transcripts in different microarray studies on hUPF1-depleted cells, for the varying efficiency of NMD on different substrates, and for the physiological role of NMD. Finally, the unified NMD model predicts the existence of a population of NMD targets that, to date, has not been appreciated and which might be clinically important.

One important aspect of future research will be to test whether the postulated mode of gene regulation by spatial rearrangement of the 3′ UTR occurs *in vivo* under normal physiological conditions, and whether it is widespread among different tissues and cell types. Furthermore, the mechanistic details of PTC recognition and the steps leading to the degradation of the mRNA are still largely unknown. More structural and biochemical data of the factors and complexes involved in NMD will be required to finally understand the molecular mechanism of NMD.

Acknowledgements

We thank Melissa Moore and Andrea Eberle for careful reading of the manuscript and for their thoughtful comments. L.S. was supported by a fellowship of the Roche Research Foundation, and O.M. is a fellow of the Max Cloëtta Foundation. The research of the laboratory is supported by the Kanton Bern and by grants to O.M. from the Swiss National Science Foundation, the Novartis Foundation for Biomedical Research, and the Helmut Horten Foundation.

References

1 Moore, M.J. (2005) From birth to death: the complex lives of eukaryotic mRNAs. *Science* 309, 1514–1518
2 Chang, Y.F. *et al.* (2007) The nonsense-mediated decay RNA surveillance pathway. *Annu. Rev. Biochem.* 76, 51–74
3 Lewis, B.P. *et al.* (2003) Evidence for the widespread coupling of alternative splicing and nonsense-mediated mRNA decay in humans. *Proc. Natl. Acad. Sci. U. S. A.* 100, 189–192
4 He, F. *et al.* (2003) Genome-wide analysis of mRNAs regulated by the nonsense-mediated and 5′ to 3′ mRNA decay pathways in yeast. *Mol. Cell* 12, 1439–1452
5 Mendell, J.T. *et al.* (2004) Nonsense surveillance regulates expression of diverse classes of mammalian transcripts and mutes genomic noise. *Nat. Genet.* 36, 1073–1078
6 Rehwinkel, J. *et al.* (2005) Nonsense-mediated mRNA decay factors act in concert to regulate common mRNA targets. *RNA* 11, 1530–1544
7 Holbrook, J.A. *et al.* (2004) Nonsense-mediated decay approaches the clinic. *Nat. Genet.* 36, 801–808
8 Ivanov, P.V. *et al.* (2008) Interactions between UPF1, eRFs, PABP and the exon junction complex suggest an integrated model for mammalian NMD pathways. *EMBO J.* 27, 736–747
9 Behm-Ansmant, I. *et al.* (2007) A conserved role for cytoplasmic poly(A)-binding protein 1 (PABPC1) in nonsense-mediated mRNA decay. *EMBO J.* 26, 1591–1601
10 Amrani, N. *et al.* (2004) A faux 3′-UTR promotes aberrant termination and triggers nonsense-mediated mRNA decay. *Nature* 432, 112–118
11 Longman, D. *et al.* (2007) Mechanistic insights and identification of two novel factors in the *C. elegans* NMD pathway. *Genes Dev.* 21, 1075–1085
12 Schwartz, A.M. *et al.* (2006) Stability of plant mRNAs depends on the length of the 3′-untranslated region. *Biochemistry (Mosc.)* 71, 1377–1384
13 Kertesz, S. *et al.* (2006) Both introns and long 3′-UTRs operate as cis-acting elements to trigger nonsense-mediated decay in plants. *Nucleic Acids Res.* 34, 6147–6157
14 Silva, A.L. *et al.* (2008) Proximity of the poly(A)-binding protein to a premature termination codon inhibits mammalian nonsense-mediated mRNA decay. *RNA* 14, 563–576
15 Eberle, B.A. *et al.* (2008) Posttranscriptional gene regulation by spatial rearrangement of the 3′ untranslated region. *PLoS Biol.* 6, e92
16 Singh, G. *et al.* (2008) A competition between stimulators and antagonists of Upf complex recruitment governs human nonsense-mediated mRNA decay. *PLoS Biol.* 6, e111
17 Behm-Ansmant, I. *et al.* (2007) mRNA quality control: an ancient machinery recognizes and degrades mRNAs with nonsense codons. *FEBS Lett.* 581, 2845–2853
18 Czaplinski, K. *et al.* (1998) The surveillance complex interacts with the translation release factors to enhance termination and degrade aberrant mRNAs. *Genes Dev.* 12, 1665–1677

19 Kashima, I. et al. (2006) Binding of a novel SMG-1-Upf1-eRF1-eRF3 complex (SURF) to the exon junction complex triggers Upf1 phosphorylation and nonsense-mediated mRNA decay. *Genes Dev.* 20, 355–367

20 Wang, W. et al. (2001) The role of Upf proteins in modulating the translation read-through of nonsense-containing transcripts. *EMBO J.* 20, 880–890

21 Kadlec, J. et al. (2006) Crystal structure of the UPF2-interacting domain of nonsense-mediated mRNA decay factor UPF1. *RNA* 12, 1817–1824

22 Anders, K.R. et al. (2003) SMG-5, required for C. elegans nonsense-mediated mRNA decay, associates with SMG-2 and protein phosphatase 2A. *EMBO J.* 22, 641–650

23 Serin, G. et al. (2001) Identification and characterization of human orthologues to Saccharomyces cerevisiae Upf2 protein and Upf3 protein (Caenorhabditis elegans SMG-4). *Mol. Cell. Biol.* 21, 209–223

24 Yamashita, A. et al. (2001) Human SMG-1, a novel phosphatidylinositol 3-kinase-related protein kinase, associates with components of the mRNA surveillance complex and is involved in the regulation of nonsense-mediated mRNA decay. *Genes Dev.* 15, 2215–2228

25 Chiu, S-Y. et al. (2003) Characterization of human Smg5/7a: A protein with similarities to Caenorhabditis elegans SMG5 and SMG7 that functions in the dephosphorylation of Upf1. *RNA* 9, 77–87

26 Ohnishi, T. et al. (2003) Phosphorylation of hUPF1 induces formation of mRNA surveillance complexes containing hSMG-5 and hSMG-7. *Mol. Cell* 12, 1187–1200

27 Fukuhara, N. et al. (2005) SMG7 is a 14-3-3-like adaptor in the nonsense-mediated mRNA decay pathway. *Mol. Cell* 17, 537–547

28 Chan, W.K. et al. (2007) An alternative branch of the nonsense-mediated decay pathway. *EMBO J.* 26, 1820–1830

29 Gehring, N.H. et al. (2005) Exon-junction complex components specify distinct routes of nonsense-mediated mRNA decay with differential cofactor requirements. *Mol. Cell* 20, 65–75

30 Bhattacharya, A. et al. (2000) Characterization of the biochemical properties of the human Upf1 gene product that is involved in nonsense-mediated mRNA decay. *RNA* 6, 1226–1235

31 Cheng, Z. et al. (2007) Structural and functional insights into the human Upf1 helicase core. *EMBO J.* 26, 253–264

32 Chamieh, H. et al. (2008) NMD factors UPF2 and UPF3 bridge UPF1 to the exon junction complex and stimulate its RNA helicase activity. *Nat. Struct. Mol. Biol.* 15, 85–93

33 Amrani, N. et al. (2006) Early nonsense: mRNA decay solves a translational problem. *Nat. Rev. Mol. Cell Biol.* 7, 415–425

34 Isken, O. and Maquat, L.E. (2007) Quality control of eukaryotic mRNA: safeguarding cells from abnormal mRNA function. *Genes Dev.* 21, 1833–1856

35 Zhang, J. (1998) Intron function in the nonsense-mediated decay of beta-globin mRNA: indications that pre-mRNA splicing in the nucleus can influence mRNA translation in the cytoplasm. *RNA* 4, 801–815

36 Buhler, M. et al. (2006) EJC-independent degradation of nonsense immunoglobulin-mu mRNA depends on 3′ UTR length. *Nat. Struct. Mol. Biol.* 13, 462–464

37 Gatfield, D. et al. (2003) Nonsense-mediated mRNA decay in Drosophila: at the intersection of the yeast and mammalian pathways. *EMBO J.* 22, 3960–3970

38 Wells, S.E. et al. (1998) Circularization of mRNA by eukaryotic translation initiation factors. *Mol. Cell* 2, 135–140

39 Hilleren, P. and Parker, R. (1999) mRNA surveillance in eukaryotes: kinetic proofreading of proper translation termination as assessed by mRNP domain organization? *RNA* 5, 711–719

40 Muhlrad, D. and Parker, R. (1999) Aberrant mRNAs with extended 3′ UTRs are substrates for rapid degradation by mRNA surveillance. *RNA* 5, 1299–1307

41 Johns, L. et al. (2007) Caenorhabditis elegans SMG-2 selectively marks mRNAs containing premature translation termination codons. *Mol. Cell. Biol.* 27, 5630–5638

42 Eulalio, A. et al. (2007) P bodies: at the crossroads of post-transcriptional pathways. *Nat. Rev. Mol. Cell Biol.* 8, 9–22

43 Sheth, U. and Parker, R. (2006) Targeting of aberrant mRNAs to cytoplasmic processing bodies. *Cell* 125, 1095–1109

44 Page, M.F. et al. (1999) SMG-2 is a phosphorylated protein required for mRNA surveillance in Caenorhabditis elegans and related to Upf1p of yeast. *Mol. Cell. Biol.* 19, 5943–5951

45 Parker, R. and Song, H. (2004) The enzymes and control of eukaryotic mRNA turnover. *Nat. Struct. Mol. Biol.* 11, 121–127

46 Garneau, N.L. et al. (2007) The highways and byways of mRNA decay. *Nat. Rev. Mol. Cell Biol.* 8, 113–126

47 Lykke-Andersen, J. (2002) Identification of a human decapping complex associated with hUpf proteins in nonsense-mediated decay. *Mol. Cell. Biol.* 22, 8114–8121

48 Gatfield, D. and Izaurralde, E. (2004) Nonsense-mediated messenger RNA decay is initiated by endonucleolytic cleavage in Drosophila. *Nature* 429, 575–578

49 Unterholzner, L. and Izaurralde, E. (2004) SMG7 acts as a molecular link between mRNA surveillance and mRNA decay. *Mol. Cell* 16, 587–596

50 Silva, A.L. et al. (2006) The canonical UPF1-dependent nonsense-mediated mRNA decay is inhibited in transcripts carrying a short open reading frame independent of sequence context. *RNA* 12, 2160–2170

51 Matsuda, D. et al. (2007) Failsafe nonsense-mediated mRNA decay does not detectably target eIF4E-bound mRNA. *Nat. Struct. Mol. Biol.* 14, 974–979

52 Shibuya, T. et al. (2004) eIF4AIII binds spliced mRNA in the exon junction complex and is essential for nonsense-mediated decay. *Nat. Struct. Mol. Biol.* 11, 346–351

53 Palacios, I.M. et al. (2004) An eIF4AIII-containing complex required for mRNA localization and nonsense-mediated mRNA decay. *Nature* 427, 753–757

54 Gehring, N.H. et al. (2003) Y14 and hUpf3b form an NMD-activating complex. *Mol. Cell* 11, 939–949

55 Gonzalez, C.I. et al. (2000) The yeast hnRNP-like protein Hrp1/Nab4 marks a transcript for nonsense-mediated decay. *Mol. Cell* 5, 489–499

56 Wittmann, J. et al. (2006) hUPF2 silencing identifies physiologic substrates of mammalian nonsense-mediated mRNA decay. *Mol. Cell. Biol.* 26, 1272–1287

57 Lelivelt, M.J. and Culbertson, M.R. (1999) Yeast Upf proteins required for RNA surveillance affect global expression of the yeast transcriptome. *Mol. Cell. Biol.* 19, 6710–6719

58 Denning, G. et al. (2001) Cloning of a novel phosphatidylinositol kinase-related kinase: characterization of the human SMG-1 RNA surveillance protein. *J. Biol. Chem.* 276, 22709–22714

59 Kadlec, J. et al. (2004) The structural basis for the interaction between nonsense-mediated mRNA decay factors UPF2 and UPF3. *Nat. Struct. Mol. Biol.* 11, 330–337

60 Glavan, F. et al. (2006) Structures of the PIN domains of SMG6 and SMG5 reveal a nuclease within the mRNA surveillance complex. *EMBO J.* 25, 5117–5125

6.3.5. Paper V

Mühlemann O, Eberle AB, Stalder L, Zamudio Orozco R. Recognition and elimination of nonsense mRNA. Biochim Biophys Acta. 2008 Sep;1779(9):538-49.

Review

Recognition and elimination of nonsense mRNA

Oliver Mühlemann *, Andrea B. Eberle, Lukas Stalder, Rodolfo Zamudio Orozco

Institute of Cell Biology, University of Berne, Baltzerstrasse 4, CH-3012 Bern, Switzerland

ARTICLE INFO

Article history:
Received 3 January 2008
Received in revised form 30 May 2008
Accepted 30 June 2008
Available online 8 July 2008

Keywords:
mRNA surveillance
Nonsense-mediated mRNA decay
Premature translation
Termination
mRNA turnover
mRNA quality control

ABSTRACT

Among the different cellular surveillance mechanisms in charge to prevent production of faulty gene products, nonsense-mediated mRNA decay (NMD) represents a translation-dependent posttranscriptional process that selectively recognizes and degrades mRNAs whose open reading frame (ORF) is truncated by a premature translation termination codon (PTC, also called "nonsense codon"). In doing so, NMD protects the cell from accumulating C-terminally truncated proteins with potentially deleterious functions. Transcriptome profiling of NMD-deficient yeast, Drosophila, and human cells revealed that 3–10% of all mRNA levels are regulated (directly or indirectly) by NMD, indicating an important role of NMD in gene regulation that extends beyond quality control [J. Rehwinkel, J. Raes, E. Izaurralde, Nonsense-mediated mRNA decay: Target genes and functional diversification of effectors, Trends Biochem. Sci. 31 (2006) 639-646.[1]]. In this review, we focus on recent results from different model organisms that indicate an evolutionarily conserved mechanism for PTC identification.

© 2008 Elsevier B.V. All rights reserved.

1. Sources of PTCs

PTCs arise from mutations in the DNA, but also on the RNA level (Table 1). Many DNA mutations within a gene will truncate the ORF. In addition to nonsense mutations, i.e. base substitutions that directly generate PTCs by changing an amino acid-encoding codon into one of the three termination codons (UAA, UAG, UGA), random nucleotide insertions and deletions shift in two of three cases the reading frame, where within the next 20 codons on average a termination codon will prematurely terminate translation. Often mutations also alter splicing signals and generate alternatively spliced mRNAs, many of which contain a PTC. Overall, it is estimated that about 30% of all known disease-associated mutations generate a PTC-containing (PTC+) mRNA [2,3].

In addition to the sources described above, PTCs arise very frequently in genes belonging to the immunoglobulin superfamily (immunoglobulins, T-cell receptors) as a consequence of programmed V(D)J rearrangements during lymphocyte maturation [4]. During the joining of a V, a D (only in heavy chains), and a J fragment, non-templated nucleotides (N nucleotides) can be added by the enzyme terminal transferase and coding nucleotides from the opposite strand (P nucleotides) can be transferred to the coding strand at the junctions of the segments. The ORF can only be maintained when nucleotides are added in multiples of three (i.e. three or six) are added, but in two thirds of the rearrangements, a frameshift results in a nonproductive allele that encodes a PTC+ t1ranscript. Interestingly, it was observed that PTCs in genes belonging to the immunoglobulin superfamily cause a much stronger reduction of the steady-state mRNA level by NMD than in other genes [5–8].

On the RNA level, errors during transcription and alternative pre-mRNA splicing generate PTC+ mRNAs. Based on a misincorporation rate for RNA polymerase II in the order of 10^{-5} per nucleotide, and assuming 10^3 to 10^4 coding nucleotides in a typical gene, only 0.05% to 0.5% of all transcripts are estimated to acquire a PTC through transcription errors. In contrast, the fraction of PTC+ transcripts generated by unproductive alternative pre-mRNA splicing is much larger. Computational analysis of human EST databases revealed that among the 60%–70% of human pre-mRNAs that are alternatively spliced, 45% had at least one splice form that was predicted to be a target of NMD [9]. Thus, about one third of all human protein-coding genes produce a PTC+ mRNA, and although their exact abundance is not known, they are likely to represent a significant fraction in the pool of the initially produced mRNAs.

2. Trans-acting factors involved in NMD

The first trans-acting factors involved in NMD have been identified in genetic screens in Saccharomyces cerevisiae and Caenorhabditis elegans. In screens for translational suppressors in S. cerevisiae, mutations in the three genes UPF1, UPF2/NMD2 and UPF3 (for Up-frameshift) were found to decrease decay rates of PTC+ mRNAs and to promote readthrough of PTCs [10–13] (Table 2). Three labs had identified in independent screens in C. elegans loss-of-function mutations in seven genes called SMG1 to SMG7 (for suppressor with morphogenetic effects on genitalia) that several years later were

* Corresponding author.
E-mail address: oliver.muehlemann@izb.unibe.ch (O. Mühlemann).

1874-9399/$ – see front matter © 2008 Elsevier B.V. All rights reserved.
doi:10.1016/j.bbagrm.2008.06.012

Table 1
Features and origins of NMD targets

Aberrant mRNAs	
Problem at DNA level	
Nonsense mutations	Base substitutions that directly generate PTCs.
Insertions and deletions	Random nucleotide insertions and deletions shift the reading frame in two of three cases, resulting in a PTC.
Mutations changing splicing signals	Mutations leading to aberrant splicing often result in a frameshift.
VDJ rearrangement	The immunoglobulin superfamily represents a special class of NMD targets that undergo very efficient NMD. Two of three rearrangements of the V, D, and J segments result in a frameshift.
Problem at RNA level	
Transcription errors	Frequency low, cause premature ORF truncation in < 1% of transcripts.
Unproductive alternative splicing	45% of alternatively spliced mRNAs are predicted to be an NMD target.
Problem at translation level	
Leaky scanning	Observed only in yeast. Ribosomes scan beyond the initiator AUG and initiate at a downstream AUG in a reading frame with a PTC.
Physiological mRNAs	
Programmed translational frameshifting	Programmed +1 or −1 frameshifts lead into a PTC, if the ribosome fails to shift the reading frame properly.
mRNAs encoding selenoproteins	UGA can be recognized as codon for selenocysteine or as PTC, depending on endogenous selenium concentration.
mRNAs with uORFs	The termination codon of the uORF is likely to be interpreted as PTC, unless the mRNA harbors stabilizing elements nearby.
mRNAs with long 3′ UTRs	Observed in all eukaryotes (including plants).
mRNAs with introns in the 3′ UTR	Observed in yeast and mammals.
Transposons and retroviruses	Observed in yeast and mammals.
Bicistronic mRNAs	Observed only in yeast.
Transcribed pseudogenes	Observed only in yeast.

recognized to be defective in NMD [14–16]. Similarity searches revealed that SMG2 is homologous to yeast UPF1, SMG3 is homologous to UPF2, and SMG4 is homologous to UPF3, respectively. As genome sequencing projects progressed, orthologs of these NMD factors were identified in other eukaryotic organisms based on homology searches [17,18]. All seven factors are present in *Homo sapiens*, and *Drosophila melanogaster* has orthologs for SMG1, UPF1, UPF2, UPF3, SMG5, and SMG6, but appears to lack an ortholog for SMG7 [19–27]. It is likely that additional, yet unknown NMD factors exist in vertebrates. Notably, Longman et al. recently identified in *C. elegans* two additional proteins, called SMGL-1/hNAG and SMGL-2/hDHX34, which are required for NMD in worms and humans [28].

UPF1, UPF2, and UPF3 constitute the conserved core of the NMD system. The nucleic acid-dependent ATPase and RNA helicase UPF1 shows the highest sequence conservation among the UPF proteins in different species [10,17,29,30] and understanding its structure, functions and regulation is key to elucidate the molecular mechanism of NMD. In the conserved region, seven group I helicase motifs can be found, and the ATPase activity of UPF1 resides in two of these helicase motifs and is linked to the 5′ to 3′ helicase activity of the protein [30]. The ATPase activity is essential for NMD in all tested species [31–35] and the RNA-binding activity of UPF1 is modulated by ATP [30,36]. UPF1 localizes predominantly to the cytoplasm [22], but has the capacity to shuttle between the nucleus and the cytoplasm [32]. UPF1 associates with the translation release factors eRF1 and eRF3 and with UPF2 [26,31,37,38] (Fig. 1). UPF1 interacts with UPF2 through its cysteine–histidine-rich domain (amino acids 115–272), which forms three Zinc-binding motifs arranged into two tandem modules, resembling the RING-box and U-box domains of ubiquitin ligases [39]. Multiple serine residues in the N- and C-terminal regions of UPF1 are targets for phosphorylation [33,40]. Indeed, UPF1 activity in humans and worms is regulated by cycles of phosphorylation and dephosphorylation that depend on additional NMD factors. Phosphorylation of UPF1 is catalyzed by SMG1 and requires UPF2 and UPF3 [31,41], whereas dephosphorylation of UPF1 is mediated by SMG5, SMG6, and SMG7, which are thought to recruit protein phosphatase 2A (PP2A) [25,31,33,41–44]. Phosphorylation of human UPF1 seems to lead to a remodeling of the UPF1-containing surveillance complex. Overexpression of a SMG1 mutant deficient in its kinase activity strongly increased UPF1 co-precipitation with eRF3, suggesting that phosphorylation of UPF1 induces the dissociation of eRF3 from UPF1 [31].

UPF2, which is also a phosphoprotein [43,45], interacts with both UPF1 and UPF3, thereby serving as a bridge between the two [22,24,26,38]. The interaction between human UPF2 and UPF3b (see below) involves the last of the three middle of eIF4G-like (MIF4G) domains of UPF2 and the RNA-binding domain (RBD) of UPF3b [46]. Human UPF2 interacts with the N-terminal Zinc-finger domain of UPF1 mainly through its C-terminal region (amino acids 1084–1272) [24,26], but amino acids 94–133 from the N-terminal region also contribute to this interaction to some extent [26]. Even though the N-terminus of UPF2 contains several nuclear localization signals (NLS) and the N-terminal 120 amino acids can target a heterologous protein

Table 2
Homologous factors involved in NMD from different species

	S. cerevisiae	C. elegans	D. melanogaster	Plants	Mammals
NMD factors	UPF1	SMG2 (UPF1)	UPF1	UPF1	UPF1 (RENT1)
	UPF2 (NMD2)	SMG3 (UPF2)	UPF2	UPF2	UPF2
	UPF3	SMG4 (UPF3)	UPF3	UPF3	UPF3a, UPF3b (UPF3X)
	−	SMG1	SMG1	ND[a]	SMG1
	−	SMG5	SMG5	ND[a]	SMG5
	−	SMG6	SMG6	ND[a]	SMG6
	−	SMG7	−	ND[a]	SMG7
	−	SMGL-1	−	SMGL-1[b]	NAG (SMGL-1)
	−	SMGL-2	SMGL-2[b]	−	DHX34 (SMGL-2)
Translation termination	SUP45	T05H4.6	eRF1	ERF1	eRF1
	SUP35	H19N07.1, K07A12.4b	eRF3	ERF3	eRF3a, eRF3b
	PAB1	pab-1	pAbp	PAB	PABPC1
Exon junction complex (EJC)	No homologs	Present, not involved in NMD	Present, not involved in NMD	Present, role in NMD is still not clear	eIF4A3
					Y14
					MAGOH
					BARENTSZ (MLN51)

[a] ND = not determined.
[b] Role in NMD is not determined.

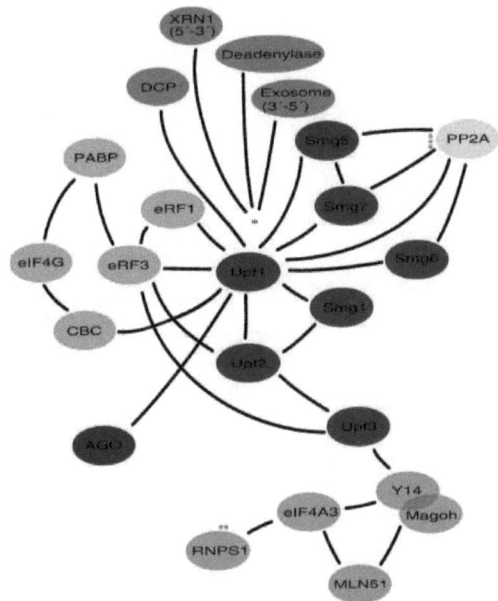

Fig. 1. Schematic illustration of NMD factors and their interactions. NMD factors are shown in red, mRNA decay factors in blue, components of the EJC in green, proteins involved in translation in orange, and RNAi factors in purple. Apart from the exceptions below, black lines depict experimental evidence for direct protein–protein interactions. *No direct physical interaction has been shown: a complex of the three UPF proteins co-immunoprecipitated with parts of the exosome, PARN (a deadenylase), XRN1, and DCP. **The interaction between RNPS1 and the EJC is not yet mapped. And it is not clear if RNPS1 directly interacts with one of the three UPF proteins. ***A complex of phosphorylated UPF1, SMG5, SMG7, and PP2A was found. But it is not clear whether SMG5 and SMG7 both interact with PP2A.

to the nucleus, UPF2 localizes at steady state to the perinuclear region of the cytoplasm [22,24,26].

UPF3 is the least conserved among the UPF proteins [17]. In contrast to yeast, worms, and flies, the human genome contains two UPF3-encoding genes, UPF3a on chromosome 13 and UPF3b on the X chromosome (also known as UPF3X), and two alternatively spliced mRNAs from each gene encode in total four UPF3 isoforms [22,26]. The shorter form of UPF3a lacks exon 4 (amino acids 141–173) and the shorter form of UPF3b lacks exon 8 (amino acids 294–306) [26]. Human UPF3 proteins contain putative NLS and NES motifs and shuttle between the nucleus and the cytoplasm, with predominant nuclear localization at steady state [22,26]. UPF3 proteins are components of the exon junction complex (EJC, see below) and bind to mRNAs through interaction with the EJC core factor Y14 [47–50].

The serine-threonine kinase SMG1 belongs to the family of phosphatidylinositol 3-kinase-related protein kinases that also includes ATM, ATR, mTOR, and DNA-PK [51]. As mentioned above, SMG1 catalyzes the phosphorylation of UPF1, and this phosphorylation is essential for NMD [19,27,31,33,40,41]. In contrast to the UPF proteins, no SMG1 ortholog has been identified in S. cerevisiae so far, and because yeast also appears to lack orthologs for SMG5, SMG6, and SMG7, it seems likely that the regulation of UPF1's phosphorylation state during NMD is limited to metazoans.

The three proteins SMG5, SMG6, and SMG7 share common structural motifs and loss-of-function mutations in any of the three genes in C. elegans lead to accumulation of phosphorylated UPF1 [33]. All three proteins contain a 14-3-3 like domain through which they bind phosphorylated UPF1 [44]. In addition, SMG5 and SMG6 both contain a C-terminal PIN domain, which is found in many proteins with ribonuclease activity. Although the crystal structure of both PIN domains is very similar, only SMG6 has the canonical triad of acidic residues that are crucial in RNase H for activity and degrades single stranded RNA in vitro, whereas SMG5 has no nuclease activity [52]. SMG5 and SMG7 interact with each other and are part of a larger complex comprising protein phosphatase 2A (PP2A) and phosphorylated UPF1 [25,42]. SMG6 also co-immunoprecipitates with PP2A and phosphorylated UPF1 [43], but this complex probably does not contain SMG5 or SMG7. The C-terminal region of SMG7 is required for the localization of SMG7, SMG5, and associated UPF1 to P-bodies, large cytoplasmic structures enriched in mRNA degradation enzymes [44,53]. In contrast, overexpressed SMG6 was detected in cytoplasmic foci different from P-bodies [53].

3. Eukaryotic release factors and poly(A) binding proteins play crucial roles in NMD

Early studies in *S. cerevisiae* already indicated that NMD is mechanistically intertwined with the process of translation termination, and more recent data indicates that NMD is activated by the intrinsically aberrant nature of premature translation termination [54]. Thus it is not surprising that several translation factors play an important role in NMD. UPF1 interacts with eRF3 and with eRF1 [31,37,55]. Both eRF1, which recognizes all three stop codons in eukaryotes and triggers peptidyl-tRNA hydrolysis, and the GTPase eRF3 are essential for translation termination [56,57]. GST pull-down experiments showed that in yeast UPF1, UPF2, UPF3, and eRF1 all interact with the GTPase domain of eRF3, whereby UPF2, UPF3, and eRF1 (but not UPF1) compete with each other for this interaction [55]. Interestingly with regard to the requirement of the ATPase/helicase activity for NMD [35,58,59], eRF1 and eRF3 were found to inhibit the RNA-dependent ATPase activity of UPF1 [37].

Another protein that plays a role in translation termination is the poly(A) binding protein (PABP). In yeast, there is only one PABP present, whereas the human genome encodes for several cytoplasmic PABPs (PABPC1, testis PABP, inducible PABP, PABP5) and a mainly nuclear form [60]. The ubiquitously expressed PABPC1 is the only PABP whose function in mRNA translation and stability has been extensively studied. PABPC1 consists of four non-identical RNA recognition motifs (RRM), of which RRM 1 and RRM 2 bind to the poly(A) tail with high affinity, a linker region, and a carboxyl terminal domain that is involved in protein–protein interactions, for example with eRF3 [61–63]. Data from yeast, flies, and humans revealed that PABP is a NMD antagonizing factor, since NMD can be suppressed by tethering PABP near the PTC [64–69].

4. Working model for an evolutionarily conserved NMD mechanism

One of the key questions about NMD is how a translation termination codon (TC) is recognized as premature and distinguished from a natural, correct TC. Although the core NMD factors are conserved from yeast to humans (see above), remarkably different models of NMD have emerged from studies of mammalian systems compared to studies in *S. cerevisiae*, *C. elegans*, *D. melanogaster*, and plants [70,71]. However, the most recent data from investigations in different organisms have begun to reveal the existence of a basic, evolutionarily conserved mechanism for PTC recognition [28,64–66,69,72,73]. It remains a challenge for the future to thoroughly scrutinize this emerging "unified NMD model" and to understand the exact molecular basis for PTC definition.

The central feature of this "unified NMD model" is that the mechanism of translation termination at a PTC is intrinsically different from translation termination at a "correct" TC. In *S. cerevisiae*, it has been demonstrated that ribosomes do not efficiently dissociate from the mRNA when stalling at a PTC, presumably because in that spatial environment they cannot receive the termination-stimulating signal from PABP [64,74]. Likewise, in flies [65] and human cells [66–69], artificial tethering of PABP into close proximity of an otherwise NMD-triggering PTC efficiently suppresses NMD. Based on reported biochemical interactions (see above), this model proposes that the decision of whether or not NMD is triggered relies on a competition between UPF1 and PABP for binding to eRF3 bound to the terminating ribosome [69] (Fig. 2). According to this model, a translation termination event is defined as "correct" if the ribosome stalls close enough from the poly(A) tail to efficiently interact with PABP, which through yet unknown mechanisms leads to fast/efficient polypeptide release and dissociation of the two ribosomal subunits (Fig. 2A). If in contrast the spatial distance between the terminating ribosome and the poly(A) tail is too big for this interaction to occur, UPF1 will bind to the ribosome-bound eRF3 instead. At this stage ("marking" step in Fig. 2B), UPF1 might still be displaced by PABP and NMD would be prevented. However, when this signal is absent, UPF2 and UPF3 will eventually bind UPF1, which is required for SMG1 to phosphorylate UPF1. UPF1 phosphorylation is believed to definitively commit the mRNA for degradation by NMD, maybe by inducing a conformational change that causes UPF1 to bind the mRNA ("licensing" step in Fig. 2B). Finally, the phosphorylated UPF1 will be bound by the 14-3-3-like phosphoserine-binding domains of SMG5, SMG6 and/or SMG7, ultimately leading to the degradation of the mRNA.

The molecular relationship between UPF1, SMG5–7 and the cellular degradation enzymes is still unclear. Evidence from yeast and mammals suggests that rapid degradation is achieved by shunting nonsense transcripts efficiently into the normal mRNA turnover pathway that proceeds through deadenylation, followed by decapping and XRN1-mediated 5′–3′ exonucleolytic decay as well as exosome-mediated 3′–5′ exonucleolytic decay [75,76] (Table 3). Consistent with this idea, several interactions between NMD and RNA decay factors have been reported. UPF1 protein co-immunoprecipitates with DCP1a and DCP2 in human cells, two subunits of the decapping complex [77]. Furthermore, UPF1, UPF2, and UPF3b have been reported to co-immunopurify with the decapping enzyme DCP2, 5′–3′ exonuclease XRN1, exosomal components PM/Scl100 (RRP6), RRP4, and RRP41, and poly(A) ribonuclease (PARN) in African green monkey cells [78]. Interestingly and in apparent contrast to the data from other species, the decay of PTC+ transcripts is initiated by endonucleolytic cleavage near the PTC in *D. melanogaster*. The resulting fragments are rapidly exonucleolytically degraded in both directions, without undergoing deadenylation or decapping [79].

5. Mammalian exon junction complex as an NMD enhancer

During pre-mRNA splicing, a protein complex called exon junction complex (EJC) is deposited about 20–24 nucleotides upstream of exon–exon junctions in spliced mRNA [80]. The four core components of the EJC Y14, MAGOH, eIF4A3, and Barentsz/MLN51 [81,82] associate with the mRNA in the nucleus and escort it to the cytoplasm. The DEAD-box adenosine triphosphatase (ATPase) eIF4A3 binds the RNA with a footprint of 7–9 nucleotides in a sequence-independent manner [47]. Additional factors associate with EJCs more dynamically and some already leave before mRNA export or only bind the EJC in the cytoplasm [80]. Identification of UPF2 and UPF3 as EJC components has immediately indicated a role for the EJC in NMD (Fig. 1). Supporting this idea, RNAi-mediated knockdown of Y14, eIF4A3 and Barentsz/MLN51 in mammalian cells reduced the NMD-mediated downregulation of PTC+ reporter mRNAs [83–86]. In contrast, studies with *D. melanogaster* and *C. elegans* showed that the EJC is dispensable for NMD in these organisms [21,28], and *S. cerevisiae* do not possess an EJC at all. Thus it appears that in organisms producing a large number of nonsense mRNAs by extensive alternative pre-mRNA splicing, the EJC may have evolved to facilitate efficient recognition and degradation of these "aberrantly spliced" transcripts.

It is believed that EJCs located within the ORF are removed from the mRNA by elongating ribosomes [87,88], and hence only EJCs located downstream of the TC would still be present when the first ribosome terminates. According to the proposed model, the presence of an EJC downstream of a terminating ribosome functions as an enhancer of NMD by facilitating SMG1/UPF2/UPF3-dependent UPF1

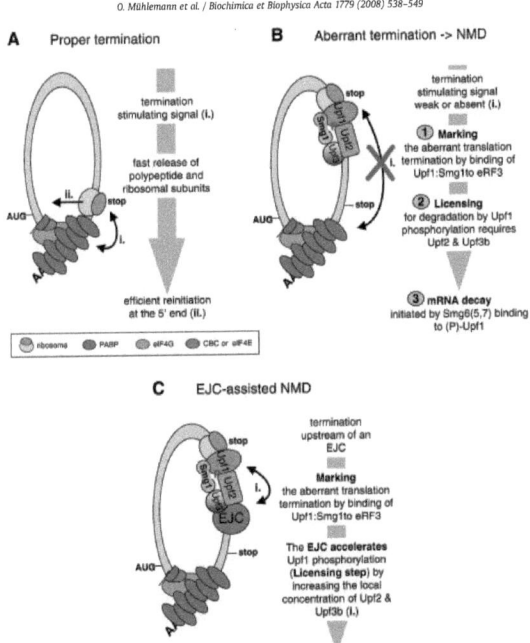

Fig. 2. Model for PTC recognition. (A) Proper translation termination depends on a stimulating signal from the poly(A) tail region and therefore only occurs when the stop codon is in spatial proximity of the poly(A) tail. (B) In the absence of this termination promoting signal, Upf1/Smg1 binds via release factors to the terminating ribosome and NMD ensues. (C) In mammals, the presence of an EJC downstream of the stop codon dramatically enhances NMD by positioning Upf2 and Upf3b ideally for promoting Upf1 phosphorylation. If the time between Upf1 binding and its phosphorylation is shortened, the competing termination promoting signal is less likely to occur, resulting in more efficient NMD. See text for details.

phosphorylation (Fig. 2C). As part of such a 3′ UTR-bound EJC, UPF2 and UPF3 are positioned ideally for immediate interaction with ribosome-bound UPF1 and SMG1. As a consequence, the time window between UPF1 binding to the terminating ribosome and its SMG1-mediated phosphorylation would be shortened and the balance between PABP action (NMD suppression) and UPF1 action (NMD promotion) would tilt towards NMD. In contrast to previously proposed mammalian NMD models, the presence of an EJC downstream of the TC is not required for PTC identification [89], but rather functions as an NMD enhancer according to this new unified NMD model.

6. Evidence for restriction of NMD to CBC-bound mRNA in mammals

However, despite of the recent convergence of NMD models derived from different model organisms, one major difference appears to exist between mammalian and yeast NMD. In mammals, several lines of evidence suggest that NMD targets exclusively cap-binding complex (CBC)-bound mRNA ([90–92], reviewed in [93,94]). CBP80, one of the two CBC constituents, interacts directly with UPF1, and this interaction was suggested to promote NMD during the so-called "pioneer round of translation" by increasing the binding of UPF1 to UPF2 [95]. In S. cerevisiae on the other hand, NMD is not restricted to CBC-bound transcripts and targets also eIF4E-bound, steady-state mRNA [96,97]. It remains to be seen if eIF4E-bound mammalian PTC+ mRNAs are immune to NMD or if they can be NMD targets under experimental conditions that prevent their previous degradation during the CBC-bound state.

7. Different phenotypes of NMD deficiency in different organisms

The phenotypes observed after inactivation of the NMD pathway vary among different organisms. In S. cerevisiae, the NMD pathway is

Table 3
Eukaryotic mRNA degradation factors

Deadenylation: removal of the poly(A) tail	
CCR4-POP2-NOT complex	Predominant deadenylase complex in yeast; the three factors are conserved in eukaryotes CCR4 and POP2: nuclease activity; NOT: accessory proteins.
PAN2/PAN3 complex	Conserved poly(A)-nuclease complex. PAN2: catalytic enzyme; PAN3 regulates and enhances PAN2 activity, and interacts with MEX67 in yeast. PAN is involved in early steps of poly(A) metabolism (in the nucleus and during export).
PARN	Poly(A)-specific exonuclease; major deadenylases in mammals (*in vitro* studies). Found in mammals, plants, and *C. elegans*, but not in *D. melanogaster* and *S. cerevisiae*.
removal of the cap structure	
DCP1-DCP2 complex	DCP2 is catalytic subunit. Decapping is inhibited by poly(A) tail and translation, and stimulated by a complex of LSM1-7 and additional proteins. DCP2 interacts with UPF1.
DCPS	Scavenger decapping enzyme. Mediates decapping of short cap oligonucleotides (e.g. produced by exosome-mediated degradation).
3′→5′ exonucleolytic degradation pathway	
XRN1	Major cytoplasmic. RNA with 5′-monophosphate (product of DCP2) is substrate for XRN1 *D. melanogaster* homologue: Pacman.
XRN2	5′-3′ exonuclease mainly involved in nuclear process. Yeast homologue: Rat1p.
3′→5′ exonucleolytic degradation pathway	
Exosome	Large ring-like protein complex containing several 3′–5′ exonucleases. Highly conserved throughout evolution, involved in RNA degradation and processing.
SKI7	Adaptor between exosome and SKI2–SKI3–SKI8 complex. Homologous to GTP-binding elongation factor 1A. Required for non-stop decay. Interacts with UPF1.
SKI2–SKI3–SKI8 complex	Central component of the 3′–5′ cytoplasmic mRNA degradation pathway. Well analyzed in yeast, but little is known about the mammalian SKI complex. SKI2 is an ATP-dependent DEAD-box RNA helicase.

Main references: Parker and Song [75] (NSMB); Wang, Lewis and Johnson (2005) (RNA); Takahashi, Araki, Sakuno and Katada (2003) (EMBO).

not essential for viability under laboratory conditions, null mutants of UPF1, UPF2, and UPF3 did not show any phenotype [54]. In *C. elegans*, mutations of SMG proteins lead to fertile worms with mild morphological defects of the genitalia [16], whereas UPF1 and UPF2 are required for larval development and cell proliferation in *D. melanogaster* [98,99]. In *Arabidopsis thaliana*, UPF1 and UPF3 are essential for the development and survival of the plant [100]. In mammals, NMD appears to be essential for viability. Homozygous Upf1 knockout mouse embryos do not survive after implantation and it was not possible to establish Upf1-/- embryonic mouse stem cells in culture [101]. However, because UPF1 is also implicated in additional cellular processes apart from NMD (see below), it is currently not clear if the lethality is indeed due to NMD inactivation. Nevertheless, we would like to point out the apparent correlation between the severity of the phenotype caused by NMD deficiency in an organism and the extent of alternative pre-mRNA splicing occurring in that organism.

8. The physiological role of NMD

Until a few years ago, NMD has been seen merely as a quality control system that rids the cell of faulty mRNAs, but recent studies indicate that NMD represents a much more sophisticated tool serving multiple purposes in gene expression. The population of NMD substrates is not only restricted to faulty transcripts, but also comprises numerous endogenous, physiological transcripts. Among those are i) mRNAs containing short upstream ORFs (uORFs,) that play a pivotal role regulating translation and turnover, ii) mRNAs encoding selenocysteine-containing proteins, iii) mRNAs harboring introns in their 3′ UTR, iv) mRNAs with long 3′ UTRs, and v) transposons and retroviruses [99,102–106]. In addition, pseudogenes, bicistronic mRNAs, and mRNAs containing signals for programmed frameshifting have been identified as NMD targets in yeast [105] (Table 1). Genome-wide transcriptome profiling in yeast, *Drosophila*, and human cells revealed that 3–10% of all mRNAs are regulated by NMD [99,104,105]. Curiously, the set of transcripts regulated by NMD varies a lot between species. And in contrast to flies, where knockdown of individual NMD factors affected essentially the same transcript population [99], the set of mRNAs regulated by UPF1 in human cells only marginally overlapped with the set of mRNAs regulated by UPF2 [104,107], which complicates interpretation of the data.

Based on our own recent data, we proposed a novel mode of translation-dependent posttranscriptional gene regulation that involves NMD ([66]; Stalder and Mühlemann, TICB, in press). According to this model, a given mRNA would be stable in a 3D configuration with the TC in physical proximity of the poly(A) tail, whereas the same mRNA could become an NMD substrate upon structural rearrangements that increase the distance between the TC and the poly(A) tail. Importantly, the protein–protein and protein–RNA interactions determining the 3D structure of the 3′ UTR can be regulated in a tissue-specific manner, during development, and by environmental cues. Along the same lines, a recent study from Giorgi et al. indicates such a translation- and NMD-dependent control of mRNA stability as a means to regulate spatially and temporally restricted protein synthesis, for example in synapses of neurons [108].

NMD is also involved in an auto-regulatory gene expression circuit termed "regulated unproductive splicing and translation" (RUST) [9]. For example, the *C. elegans* ribosomal proteins L3, L7a, L10a and L12 and the SR proteins SRp20 and SRp30b are regulated in this manner [109,110]. By alternative splicing of their pre-mRNA, both a productive mRNA encoding the full-length protein and a nonproductive PTC+ mRNA are produced. At a certain concentration, the protein starts to alter the splicing of its own pre-mRNA towards production of more nonproductive PTC+ mRNA, which is rapidly degraded by NMD. In mammals, the SR protein SC35 has also been shown to auto-regulate its own expression using RUST. At high levels of SC35 protein, it binds its own pre-mRNA, inducing the production of alternatively spliced PTC+ mRNA [111]. Moreover, NMD can act in concert with alternative pre-mRNA splicing and miRNA silencing to regulate gene expression. In neuronal cells, miR-124 regulates the expression of PTBP1, which among others targets PTBP2 for alternative splicing, resulting in a PTC-containing PTBP2 mRNA. Upon expression of miR-124, PTBP1 protein levels are reduced, leading to an accumulation of correctly spliced PTBP2 mRNA and the production of PTBP2 protein [112]. In *A. thaliana*, the clock-regulated RNA-binding protein AtGRP7 regulates its abundance by an auto-regulatory negative feedback loop in a posttranscriptional manner. AtGRP7 targets its own pre-mRNA, leading to the production of alternatively spliced PTC+ transcripts [113]. Thus, the coupling of NMD with alternative splicing seems to be widespread among eukaryotes. Since NMD rarely downregulates a PTC+ transcript to less than 10%, physiological relevant levels of truncated proteins may still be produced from the remaining PTC+ transcript [114]. For example, the steady-state level of the PTC+ mRNA encoding the FcεRIβ receptor is very low, but nonetheless the truncated protein competes effectively with the full-length protein [115].

Interestingly, the overall NMD capacity can vary considerably between different cell lines, tissues, and individuals [116–121]. It has been proposed that these variations of the NMD efficiency might contribute to the phenotypic variability of hereditary disorders [2,122]. Recently, variability in NMD efficiency between cell lines has been reported to correlate with intracellular RNPS1 concentrations [123].

9. Involvement of NMD factors in additional cellular processes

Given the intimate coupling of NMD to translation, it may not come as a surprise that NMD factors have been found to play a role in various aspects of translation. UPF1, UPF2, UPF3b, and the EJC factors Y14, MAGOH, and RNPS1 have been reported to stimulate translation [124]. These proteins enhanced translational yield and polysome association when tethered inside the ORF of an intronless reporter mRNA, but the mechanism underlying this effect remains to be elucidated. In *S. cerevisiae*, NMD factors were shown to increase the fidelity of translation termination. Loss-of-function mutations in UPF1, UPF2 or UPF3 cause nonsense suppression, resulting in production of C-terminally extended proteins [55,125]. The function of UPF1-3 in nonsense suppression might be attributed to their interaction with eRF1 and eRF3. However, knockdown of UPF1 and UPF2 in mammalian cells had no effect on nonsense suppression [32], and it remains to be shown if NMD factors are involved in nonsense suppression in other species than yeast.

More surprising was the finding that NMD factors are implicated in telomere maintenance. Est1p is involved in telomerase regulation in yeast [126]. Three Est1p-like proteins, EST1A, EST1B, and EST1C have been identified in humans and were found to be the homologs of *C. elegans* SMG6, SMG5, and SMG7, respectively [21,25,43,127,128]. Human SMG6 (hEST1A) associates with telomerase activity in HeLa cell extracts and overexpression of this protein leads to accumulation of chromosome-end fusions [127]. Very recently, transcripts originating from telomeric repeat regions have been discovered in mammalian cells [129]. Intriguingly, UPF1, UPF2, SMG1, and SMG6 were all found to be enriched at telomeres, and they negatively regulate the association of the telomeric repeat-containing RNAs (TERRAs) with chromatin [129]. Thus, these NMD factors provide a molecular link between TERRA regulation and the maintenance of telomere integrity and therewith genome stability.

Furthermore and like ATM, another PI-3 kinase-related kinase, human SMG1 phosphorylates the cell-cycle checkpoint protein p53 [130]. SMG1 is required for optimal p53 activation upon genotoxic stress and SMG1 depletion causes spontaneous DNA damage and increased sensitivity to ionizing radiation [130]. UPF1 also gets phosphorylated under such conditions by ATR and associates with chromatin in discrete spots [131]. UPF1 depletion leads to a cell-cycle arrest in early S-phase and an ATR-dependent DNA-damage response [131], which together with the identification of UPF1 as the DNA polymerase δ -associated helicase [132] strongly suggests a function for UPF1 in DNA repair and corroborates a function for NMD factors in safeguarding genome stability.

Moreover, studies in *C. elegans* revealed that *SMG-2* (*C. elegans* UPF1), *SMG-5*, and *SMG-6* are required for establishing persistent RNAi-mediated gene silencing. Target genes could initially be silenced by dsRNA in mutant *smg2*, *smg5*, and *smg6* animals, but the target mRNA levels recovered within 4days, whereas the knockdown persisted in wild-type worms [133]. *Smg-3* and *smg-4* mutants gave a similar, but weaker phenotype, whereas RNAi persistence was not affected in *smg-1* mutants. Thus, the fast recovery from RNAi does not generally depend on inactivation of NMD. In addition to this genetic link between NMD and RNAi, a physical link between NMD and RNAi factors was recently reported: UPF1 co-immunoprecipitates AGO1 and AGO2 in an RNAse A insensitive manner, and UPF1 was identified in AGO1-associated mRNPs [134].

10. Subcellular localization of NMD

A recurring topic in the NMD field has been the debate about the subcellular location of NMD in mammalian cells. Although it is undisputed that NMD depends on translation, a process that generally is thought to be confined to the cytoplasm, most examined nonsense mRNAs in mammalian cells are found to be degraded when still physically attached to the nucleus (i.e. in the RNA isolated from purified nuclei). Models to resolve this apparent paradox range from proposing translation by cytoplasmic ribosomes associated with the outer nuclear membrane to postulating nuclear scanning of the reading frame at the site of transcription by a yet unknown nuclear frame reader (reviewed in [135,136]). The reported evidence for nuclear translation in human cells [137], the confinement of NMD to CBC-bound mRNA [90], and finding that NMD was not affected by inhibition of mRNA export [138] are consistent with the hypothesis of intranuclear NMD. On the other hand, Lykke-Andersen et al. showed very recently that expression of polypeptides designed to inhibit various interactions between NMD factors specifically inhibited NMD when expressed exogenously in the cytoplasm [139]. The same polypeptides did however not inhibit NMD when they were confined to the nucleus by addition of a nuclear localization signal (NLS), suggesting that most if not all NMD occurs in the cytoplasm of mammalian cells [139].

For *S. cerevisiae*, evidence is accumulating that NMD occurs in the cytoplasmic processing bodies (P-bodies) [140], and nonsense mRNAs in mammalian cells were also proposed to be degraded in similar cytoplasmic foci that are enriched in decapping and degradation enzymes, NMD factors, and effectors of the RNAi silencing pathway (reviewed in [141]). SMG7 might provide the molecular link between NMD and the degradation machinery in mammalian cells, through its 14-3-3 like domain in the N-terminus that interacts with phosphorylated UPF1 and SMG5. When overexpressed, SMG7 accumulates together with SMG5 and UPF1 in P-bodies. SMG7 elicits RNA degradation independently of a PTC when tethered to a reporter mRNA [53].

11. Nuclear aspects of NMD

Regardless of the controversy about nuclear versus cytoplasmic NMD, examples where the presence of PTC in the transcript affects nuclear processes such as pre-mRNA splicing and transcription have been reported. The crucial question with the examples of nonsense-associated altered splicing (NAS) is, if splicing is altered specifically as a consequence of the ORF truncation, or if the PTC-causing mutation affects splicing directly. In many examples of NAS that were thoroughly investigated, it was found that not only nonsense, but also some missense and silent mutations caused alternative splicing, and that these mutations disrupted exonic splicing enhancers critical for exon inclusion [142–147]. However, evidence for frame-dependent NAS has also been reported [148–152]. The requirement for UPF1 is a distinction criteria for this second type of NAS that is believed to be specific for frame-disrupting mutations, whereas other NMD factors do not appear to be required [32,152]. Interestingly, specific NMD-inactivating amino acid substitutions in UPF1 did still support frame-dependent NAS, indicating genetically separable functions of UPF1 in these two processes [32]. That recognition of a PTC in an mRNA could generate a signal that alters splice site selection on its own pre-mRNA species is intriguing and somewhat paradoxical, and the underlying mechanism remains to be elucidated.

Yet another unexpected PTC-dependent nuclear effect was recently discovered in Ig-α and Ig-γ minigenes [153]. When these minigenes were stably integrated into the genome of HeLa cells, transcription of PTC-containing genes was silenced. This transcriptional silencing is PTC-specific and accompanied by chromatin remodeling, manifested by the loss of typical marks for transcriptionally active euchromatin (acetylated histone H3) and a concomitant accumulation of heterochromatin marks (H3K9 methylation). Consistently, this nonsense-mediated transcriptional gene silencing (NMTGS) can be reversed by treatment of the cells with histone deacetylase inhibitors. Remarkably, NMTGS was inhibited by overexpression of the siRNAse 3′hExo, which suggests that small interfering RNAs (siRNAs) and hence the RNA interference (RNAi) pathway are involved [153]. However, attempts to detect Ig-α-specific siRNAs have not been successful until now (O.M., unpublished results). The NMTGS pathway branches from the NMD pathway after translation of the PTC-containing mRNA and UPF1-dependent PTC recognition [154]. So far, NMTGS was only observed with PTC-containing Ig-α and Ig-γ minigenes, but not with an Ig-κ minigene or other classical NMD reporter genes [153], which led us to speculate that NMTGS might be important to silence expression of non-productively rearranged heavy chain alleles in B cells, thereby constituting the second half of the allelic exclusion system. However, investigation of clonal lines of immortalized murine pro-B cells did not reveal a difference of the transcriptional state between the productively and the non-productively rearranged heavy chain allele (A.B.E. and O.M., unpublished results). Thus, until now the biological significance of NMTGS remains enigmatic.

12. Clinical aspects of NMD and therapeutic approaches

It is estimated that approximately 30% of all human genetic diseases arise as a consequence of nonsense or frameshift mutations that disrupt the ORF of genes by the introduction of PTCs [122]. Expression of these mutant genes lead to the synthesis of nonsense mRNAs that if translated conduce to the production of truncated proteins with residual or no function, or in some cases with gain-of-function or dominant-negative activities (reviewed in [2,122,155]; see Table 4). In most cases truncated proteins are not produced, because NMD detects and rapidly destabilize the PTC-containing mRNAs. By doing so, NMD acts as a "silencer" of PTC-containing genes and protect cells against the accumulation of potentially toxic C-terminally truncated proteins. Such a protective role of NMD has been documented from studying β-thalassemic disorders, in which production of β-globin polypeptides and consequently of hemoglobin is compromised [156]. Many β thalassemias are caused by the introduction of PTCs in the β-globin gene. In most patients these diseases show a recessive mode of inheritance, but there are also a few cases in which the mode of inheritance is dominant. PTCs that trigger NMD and hence lead to no or very little production of truncated β-globin are associated with the recessive mode of inheritance, but if the PTC is localized in the last exon of the β-globin gene, the transcripts produced are NMD-insensitive and are translated into truncated proteins with dominant-negative effects. PTC-generating mutations in the gene SOX10, which encodes a transcription factor essential for the development of cells in the neural crest lineage, provide another example of the protecting role of NMD against production of truncated toxic proteins [157]. Different nonsense mutations in SOX10 generate one of two distinct neurological diseases. The more severe one is a complex neurocristopathy called PCWH that is caused by the presence of PTCs in exon 5, the last exon of the gene. The less severe neurocristopathy (termed WS4) has a more restricted phenotype and PTCs in patients with WS4 reside principally in exons 3 and 4. Whereas SOX10 mRNAs with PTCs in exon 3 or 4 are degraded by NMD, resulting in the virtual absence of truncated SOX10 protein and haploinsufficiency as a consequence, PCWH is the consequence of the activities of dominant-negative truncated proteins produced by the translation of NMD-insensitive SOX10 mRNAs with PTCs in the last exon.

Table 4
Examples of genetic diseases where NMD modulates the phenotype

Gene name	NMD efficiency	Effect of mutation/phenotype
NMD beneficial		
β-globin	High	Heterozygotes healthy; recessively inherited β-thalassemia major
	Low	Dominantly inherited β-thalassemia intermedia
SOX10	High	Haploinsufficiency leads to less severe neurocristopathy (WS4: Waardenburg and Hirschsprung diseases)
	Low	Dominantly inherited complex neurocristopathy (PCWH: peripheral demyelinating neuropathy, central dysmyelinating leukodystrophy, Waardenburg syndrome and Hirschsprung disease)
Rhodopsin	High	Heterozygotes have abnormalities on retinogram, but no clinical disease; recessively inherited blindness
	Low	Dominantly inherited blindness
Receptor tyrosine kinase-like orphan receptor 2	High	Heterozygotes healthy; recessively inherited Robinow syndrome (orodental abnormalities, hypoplastic genitalia, multiple rib/vertebral anomalies)
	Low	Dominantly inherited brachydactyly type B (shortening of digits and metacarpals)
Cone-rod homeobox	High	Mutation found in unaffected heterozygotes (no homozygotes found)
	Low	Dominantly inherited retinal disease
Coagulation factor X	High	Heterozygotes healthy; recessively inherited bleeding tendency
	Low	Dominantly inherited bleeding tendency
NMD detrimental		
Dystrophin	High	Severe form of muscular dystrophy (Duchenne muscular dystrophy)
	Low	Milder form of muscular dystrophy (Becker muscular dystrophy)
CFTR (cystic fibrosis)	High	Severe cystic fibrosis
	Low	Milder form of cystic fibrosis

Adapted from [2].

In contrast to the two examples above, where the beneficial effect of NMD is evident, there are also cases in which NMD aggravates the disease phenotype, because the destruction of nonsense mRNAs prevents sufficient accumulation of truncated, yet still functional proteins. Mutant forms of the dystrophin gene provide an example of this detrimental NMD effect. PTCs in regions of the dystrophin gene where they trigger NMD are associated with the severe form of the disorder called Duchenne muscular dystrophy (DMD), while the milder form, the so-called Becker muscular dystrophy (BMD), is associated with nonsense mRNAs that are not recognized by NMD and hence serve as templates for synthesis of C-terminally truncated dystrophin protein [117].

For many genetic disorders caused by PTC-generating mutations, there are no effective treatments available. Because NMD plays an important role in modulating the clinical manifestations of such diseases, interfering with NMD represents a promising therapeutic strategy. For those cases where the prematurely truncated protein is still functional, inhibiting rapid degradation of the nonsense mRNA would in principle suffice to elevate the protein concentration and ameliorate the condition of patients. But in most cases, production of the full-length protein would be necessary to restore function, which can be achieved by promoting readthrough of the PTC

[158,159]. The first promising results were obtained with aminoglycoside antibiotics, which at very high concentrations promote the readtrough of stop codons in eukaryotic mRNAs [158]. The aminoglycoside gentamicin suppresses stop codons in *in vitro* assays, and beneficial effects of gentamicin treatment were reported from clinical trials with cystic fibrosis patients [160]. However, the high gentamicin concentrations also had toxic side effects on kidneys and ears, questioning the usefulness of its systemic application. More recently, a new readthrough-promoting compound with no structural relationship to aminoglycosides has been developed. Remarkably, this small organic molecule called PTC124 selectively induces readthrough of premature but not normal termination codons [161]. PTC124 rescued striated muscle function in *mdx* mice expressing dystrophin nonsense alleles and the drug is currently being tested in clinical phase II trials on patients suffering from cystic fibrosis or DMD.

13. Open questions – future goals

Despite of intensive research and substantial progress during recent years, several fundamental questions about NMD are still waiting to be solved. Investigation of the underlying molecular mechanism would be greatly facilitated if an *in vitro* system could be developed that recapitulates the main features of NMD. To understand the NMD mechanism, understanding in more detail the regulation and function of the key NMD factor UPF1 will be crucial. Still very little is also known about the molecular link between recognition of an aberrant translation termination event and the rapid degradation of this mRNA. And equally little is known about the actual degradation mechanism of nonsense mRNA. With regard to the various seemingly unrelated cellular processes that NMD factors affect, it will be interesting to learn, if these processes are somehow mechanistically coupled or if the NMD factors have a double life with two totally independent jobs. Furthermore, the role of NMD as a translation-dependent posttranscriptional regulator of many physiological mRNAs that are NMD targets needs to be explored. And finally, a detailed understanding of the mechanisms of PTC recognition and nonsense mRNA degradation will benefit the development of highly specific tools for manipulating NMD and therewith gene expression. Such tools are not only promising for future therapies of diseases caused by PTC-generating mutations, but would also offer various applications in biotechnology.

Acknowledgments

O.M. is a fellow of the Max Cloëtta Foundation, and R.Z.O. is supported by a fellowship from CONACYT, México. The research of the authors is supported by grants to O.M. from the Swiss National Science Foundation, the Novartis Foundation for Biomedical Research and the Helmut Horten Foundation, and by the Kanton Bern.

References

[1] J. Rehwinkel, J. Raes, E. Izaurralde, Nonsense-mediated mRNA decay: target genes and functional diversification of effectors, Trends Biochem. Sci. 31 (2006) 639–646.
[2] J.A. Holbrook, G. Neu-Yilik, M.W. Hentze, A.E. Kulozik, Nonsense-mediated decay approaches the clinic, Nat. Genet. 36 (2004) 801–808.
[3] J.T. Mendell, H.C. Dietz, When the message goes awry. Disease-producing mutations that influence mRNA content and performance, Cell 107 (2001) 411–414.
[4] S. Li, M.F. Wilkinson, Nonsense surveillance in lymphocytes? Immunity 8 (1998) 135–141.
[5] A. Connor, E. Wiersma, M.J. Shulman, On the linkage between RNA processing and RNA translatability, J. Biol. Chem. 269 (1994) 25178–25184.
[6] H.M. Jack, J. Berg, M. Wabl, Translation affects immunoglobulin mRNA stability, Eur. J. Immunol. 19 (1989) 843–847.
[7] M.S. Carter, S. Li, M.F. Wilkinson, A splicing-dependent regulatory mechanism that detects translation signals, EMBO. J. 15 (1996) 5965–5975.
[8] A. Buzina, M.J. Shulman, Infrequent translation of a nonsense codon is sufficient to decrease mRNA level, Mol. Biol. Cell 10 (1999) 515–524.
[9] B.P. Lewis, R.E. Green, S.E. Brenner, Evidence for the widespread coupling of alternative splicing and nonsense-mediated mRNA decay in humans, Proc. Natl. Acad. Sci. U. S. A. 100 (2003) 189–192.
[10] P. Leeds, S.W. Peltz, A. Jacobson, M.R. Culbertson, The product of the yeast UPF1 gene is required for rapid turnover of mRNAs containing a premature translational termination codon, Genes. Dev. 5 (1991) 2303–2314.
[11] P. Leeds, J.M. Wood, B.S. Lee, M.R. Culbertson, Gene products that promote mRNA turnover in *Saccharomyces cerevisiae*, Mol. Cell Biol. 12 (1992) 2165–2177.
[12] F. He, A. Jacobson, Identification of a novel component of the nonsense-mediated mRNA decay pathway by use of an interacting protein screen, Genes Dev. 9 (1995) 437–454.
[13] Y. Cui, K.W. Hagan, S. Zhang, S.W. Peltz, Identification and characterization of genes that are required for the accelerated degradation of mRNAs containing a premature translational termination codon, Genes Dev. 9 (1995) 423–436.
[14] B.M. Cali, S.L. Kuchma, J. Latham, P. Anderson, smg-7 is required for mRNA surveillance in *Caenorhabditis elegans*, Genetics 151 (1999) 605–616.
[15] J. Hodgkin, A. Papp, R. Pulak, V. Ambros, P. Anderson, A new kind of informational suppression in the nematode *Caenorhabditis elegans*, Genetics 123 (1989) 301–313.
[16] R. Pulak, P. Anderson, mRNA surveillance by the *Caenorhabditis elegans* smg genes, Genes. Dev. 7 (1993) 1885–1897.
[17] M.R. Culbertson, P.F. Leeds, Looking at mRNA decay pathways through the window of molecular evolution, Curr. Opin. Genet. Dev. 13 (2003) 207–214.
[18] L.E. Maquat, Nonsense-mediated mRNA decay: a comparative analysis of different species, Curr. Genomics 5 (2004) 175–190.
[19] G. Denning, L. Jamieson, L.E. Maquat, E.A. Thompson, A.P. Fields, Cloning of a novel phosphatidylinositol kinase-related kinase: characterization of the human SMG-1 RNA surveillance protein, J. Biol. Chem. 276 (2001) 22709–22714.
[20] S.E. Applequist, M. Selg, C. Raman, H.M. Jack, Cloning and characterization of HUPF1, a human homolog of the *Saccharomyces cerevisiae* nonsense mRNA-reducing UPF1 protein, Nucleic Acids Res. 25 (1997) 814–821.
[21] D. Gatfield, L. Unterholzner, F.D. Ciccarelli, P. Bork, E. Izaurralde, Nonsense-mediated mRNA decay in *Drosophila*: at the intersection of the yeast and mammalian pathways, EMBO J. 22 (2003) 3960–3970.
[22] J. Lykke-Andersen, M.D. Shu, J.A. Steitz, Human Upf proteins target an mRNA for nonsense-mediated decay when bound downstream of a termination codon, Cell 103 (2000) 1121–1131.
[23] H.A. Perlick, S.M. Medghalchi, F.A. Spencer, R.J. Kendzior, Jr., H.C. Dietz, Mammalian orthologues of a yeast regulator of nonsense transcript stability, Proc. Natl. Acad. Sci. U. S. A. 93 (1996) 10928–10932.
[24] J.T. Mendell, S.M. Medghalchi, R.G. Lake, E.N. Noensie, H.C. Dietz, Novel Upf2p orthologues suggest a functional link between translation initiation and nonsense surveillance complexes, Mol. Cell Biol. 20 (2000) 8944–8957.
[25] T. Ohnishi, A. Yamashita, I. Kashima, T. Schell, K.R. Anders, A. Grimson, T. Hachiya, M.W. Hentze, P. Anderson, S. Ohno, Phosphorylation of hUPF1 induces formation of mRNA surveillance complexes containing hSMG-5 and hSMG-7, Mol. Cell 12 (2003) 1187–1200.
[26] G. Serin, A. Gersappe, J.D. Black, R. Aronoff, L.E. Maquat, Identification and characterization of human orthologues to *Saccharomyces cerevisiae* Upf2 protein and Upf3 protein (*Caenorhabditis elegans* SMG-4), Mol. Cell Biol. 21 (2001) 209–223.
[27] A. Yamashita, T. Ohnishi, I. Kashima, Y. Taya, S. Ohno, Human SMG-1, a novel phosphatidylinositol 3-kinase-related protein kinase, associates with components of the mRNA surveillance complex and is involved in the regulation of nonsense-mediated mRNA decay, Genes Dev. 15 (2001) 2215–2228.
[28] D. Longman, R.H. Plasterk, I.L. Johnstone, J.F. Caceres, Mechanistic insights and identification of two novel factors in the *C. elegans* NMD pathway, Genes Dev. (2007).
[29] K. Czaplinski, Y. Weng, K.W. Hagan, S.W. Peltz, Purification and characterization of the Upf1 protein: a factor involved in translation and mRNA degradation, RNA 1 (1995) 610–623.
[30] A. Bhattacharya, K. Czaplinski, P. Trifillis, F. He, A. Jacobson, S.W. Peltz, Characterization of the biochemical properties of the human Upf1 gene product that is involved in nonsense-mediated mRNA decay, RNA 6 (2000) 1226–1235.
[31] I. Kashima, A. Yamashita, N. Izumi, N. Kataoka, R. Morishita, S. Hoshino, M. Ohno, G. Dreyfuss, S. Ohno, Binding of a novel SMG-1-Upf1-eRF1-eRF3 complex (SURF) to the exon junction complex triggers Upf1 phosphorylation and nonsense-mediated mRNA decay, Genes Dev. 20 (2006) 355–367.
[32] J.T. Mendell, C.M. ap Rhys, H.C. Dietz, Separable roles for rent1/hUpf1 in altered splicing and decay of nonsense transcripts, Science 298 (2002) 419–422.
[33] M.F. Page, B. Carr, K.R. Anders, A. Grimson, P. Anderson, SMG-2 is a phosphorylated protein required for mRNA surveillance in *Caenorhabditis elegans* and related to Upf1p of yeast, Mol. Cell Biol. 19 (1999) 5943–5951.
[34] X. Sun, H.A. Perlick, H.C. Dietz, L.E. Maquat, A mutated human homologue to yeast Upf1 protein has a dominant-negative effect on the decay of nonsense-containing mRNAs in mammalian cells, Proc. Natl. Acad. Sci. U. S. A. 95 (1998) 10009–10014.
[35] Y. Weng, K. Czaplinski, S.W. Peltz, Genetic and biochemical characterization of mutations in the ATPase and helicase regions of the Upf1 protein, Mol. Cell Biol. 16 (1996) 5477–5490.
[36] Z. Cheng, D. Muhlrad, M.K. Lim, R. Parker, H. Song, Structural and functional insights into the human Upf1 helicase core, EMBO J. 26 (2007) 253–264.
[37] K. Czaplinski, M.J. Ruiz-Echevarria, S.V. Paushkin, X. Han, Y. Weng, H.A. Perlick, H.C. Dietz, M.D. Ter-Avanesyan, S.W. Peltz, The surveillance complex interacts with the translation release factors to enhance termination and degrade aberrant mRNAs, Genes Dev. 12 (1998) 1665–1677.

[38] F. He, A.H. Brown, A. Jacobson, Upf1p, Nmd2p, and Upf3p are interacting components of the yeast nonsense-mediated mRNA decay pathway, Mol. Cell Biol. 17 (1997) 1580–1594.
[39] J. Kadlec, D. Guilligay, R.B. Ravelli, S. Cusack, Crystal structure of the UPF2-interacting domain of nonsense-mediated mRNA decay factor UPF1, RNA 12 (2006) 1817–1824.
[40] M. Pal, Y. Ishigaki, E. Nagy, L.E. Maquat, Evidence that phosphorylation of human Upf protein varies with intracellular location and is mediated by a wortmannin-sensitive and rapamycin-sensitive PI 3-kinase-related kinase signaling pathway, RNA 7 (2001) 5–15.
[41] A. Grimson, S. O'Connor, C.L. Newman, P. Anderson, SMG-1 is a phosphatidyli-nositol kinase-related protein kinase required for nonsense-mediated mRNA Decay in Caenorhabditis elegans, Mol. Cell Biol. 24 (2004) 7483–7490.
[42] K.R. Anders, A. Grimson, P. Anderson, SMG-5, required for C. elegans nonsense-mediated mRNA decay, associates with SMG-2 and protein phosphatase 2A, EMBO J. 22 (2003) 641–650.
[43] S.Y. Chiu, E. Serin, O. Ohara, L.E. Maquat, Characterization of human Smg5/7a: A protein with similarities to Caenorhabditis elegans SMG5 and SMG7 that functions in the dephosphorylation of Upf1, RNA 9 (2003) 77–87.
[44] N. Fukuhara, J. Ebert, L. Unterholzner, D. Lindner, E. Izaurralde, E. Conti, SMG7 is a 14-3-3-like adaptor in the nonsense-mediated mRNA decay pathway, Mol. Cell 17 (2005) 537–547.
[45] W. Wang, I.J. Cajigas, S.W. Peltz, M.F. Wilkinson, C.I. Gonzalez, Role for Upf2p phosphorylation in Saccharomyces cerevisiae nonsense-mediated mRNA decay, Mol. Cell Biol. 26 (2006) 3390–3400.
[46] J. Kadlec, E. Izaurralde, S. Cusack, The structural basis for the interaction between the nonsense-mediated mRNA decay factors UPF2 and UPF3, Nat. Struct. Mol. Biol. 11 (2004) 330–337.
[47] H. Le Hir, E. Izaurralde, L.E. Maquat, M.J. Moore, The spliceosome deposits multiple proteins 20–24 nucleotides upstream of mRNA exon-exon junctions, EMBO J. 19 (2000) 6860–6869.
[48] H. Le Hir, D. Gatfield, E. Izaurralde, M.J. Moore, The exon-exon junction complex provides a binding platform for factors involved in mRNA export and nonsense-mediated mRNA decay, EMBO J. 20 (2001) 4987–4997.
[49] V.N. Kim, N. Kataoka, G. Dreyfuss, Role of the nonsense-mediated decay factor hUpf3 in the splicing-dependent exon-exon junction complex, Science 293 (2001) 1832–1836.
[50] V.N. Kim, J. Yong, N. Kataoka, L. Abel, M.D. Diem, G. Dreyfuss, The Y14 protein communicates to the cytoplasm the position of exon-exon junctions, EMBO J. 20 (2001) 2062–2068.
[51] R.T. Abraham, PI 3-kinase related kinases: 'big' players in stress-induced signaling pathways, DNA Rep. (Amst). 3 (2004) 883–887.
[52] F. Glavan, I. Behm-Ansmant, E. Izaurralde, E. Conti, Structures of the PIN domains of SMG6 and SMG5 reveal a nuclease within the mRNA surveillance complex, EMBO J. 25 (2006) 5117–5125.
[53] L. Unterholzner, E. Izaurralde, SMG7 acts as a molecular link between mRNA surveillance and mRNA decay, Mol. Cell 16 (2004) 587–596.
[54] N. Amrani, M.S. Sachs, A. Jacobson, Early nonsense: mRNA decay solves a translational problem, Nat. Rev. Mol. Cell Biol. 7 (2006) 415–425.
[55] W. Wang, K. Czaplinski, Y. Rao, S.W. Peltz, The role of Upf proteins in modulating the translation read-through of nonsense-containing transcripts, EMBO J. 20 (2001) 880–890.
[56] L. Kisselev, M. Ehrenberg, L. Frolova, Termination of translation: interplay of mRNA, rRNAs and release factors?, EMBO J. 22 (2003) 175–182.
[57] V.A. Mitkevich, A.V. Kononenko, I.Y. Petrushanko, D.V. Yanvarev, A.A. Makarov, L.L. Kisselev, Termination of translation in eukaryotes is mediated by the quaternary eRF1*eRF3*GTP*Mg2+ complex. The biological roles of eRF3 and prokaryotic RF3 are profoundly distinct, Nucleic Acids Res. 34 (2006) 3947–3954.
[58] Y. Weng, K. Czaplinski, S.W. Peltz, Identification and characterization of mutations in the UPF1 gene that affect nonsense suppression and the formation of the Upf protein complex but not mRNA turnover, Mol. Cell Biol. 16 (1996) 5491–5506.
[59] Y.M. Weng, M.J. Ruiz-Echevarria, S. Zhang, Y. Cui, K. Czaplinski, J.D. Dinman, S.W. Peltz, Characterization of the nonsense-mediated mRNA decay pathway and its effect on modulating translation termination and programmed frameshifting, mRNA Metabolism and Post-Transcriptional Gene Regulation, vol. 17, Wiley-Liss Inc, New York, 1997, pp. 241–263.
[60] U. Kuhn, E. Wahle, Structure and function of poly(A) binding proteins, Biochim. Biophys. Acta 1678 (2004) 67–84.
[61] B. Cosson, N. Berkova, A. Couturier, S. Chabelskaya, M. Philippe, G. Zhouravleva, Poly(A)-binding protein and eRF3 are associated in vivo in human and Xenopus cells, Biol. Cell 94 (2002) 205–216.
[62] S. Hoshino, M. Imai, T. Kobayashi, N. Uchida, T. Katada, The eukaryotic polypeptide chain releasing factor (eRF3/GSPT) carrying the translation termination signal to the 3′-Poly(A) tail of mRNA. Direct association of erf3/GSPT with polyadenylate-binding protein, J. Biol. Chem. 274 (1999) 16677–16680.
[63] G. Kozlov, G. De Crescenzo, N.S. Lim, N. Siddiqui, D. Fantus, A. Kahvejian, J.F. Trempe, D. Elias, I. Ekiel, N. Sonenberg, M. O'Connor-McCourt, K. Gehring, Structural basis of ligand recognition by PABC, a highly specific peptide-binding domain found in poly(A)-binding protein and a HECT ubiquitin ligase, EMBO J. 23 (2004) 272–281.
[64] N. Amrani, R. Ganesan, S. Kervestin, D.A. Mangus, S. Ghosh, A. Jacobson, A faux 3′-UTR promotes aberrant termination and triggers nonsense-mediated mRNA decay, Nature 432 (2004) 112–118.
[65] I. Behm-Ansmant, D. Gatfield, J. Rehwinkel, V. Hilgers, E. Izaurralde, A conserved role for cytoplasmic poly(A)-binding protein 1 (PABPC1) in nonsense-mediated mRNA decay, EMBO J. (2007).
[66] A.B. Eberle, L. Stalder, H. Mathys, R. Zamudio Orozco, O. Muhlemann, Post-transcriptional gene regulation by spatial rearrangement of the 3′ untranslated region, PLoS Biol. 6 (2008) e92.
[67] P.V. Ivanov, N.H. Gehring, J.B. Kunz, M.W. Hentze, A.E. Kulozik, Interactions between UPF1, eRFs, PABP and the exon junction complex suggest an integrated model for mammalian NMD pathways, EMBO J. (2008).
[68] A.L. Silva, P. Ribeiro, A. Inacio, S.A. Liebhaber, L. Romao, Proximity of the poly(A)-binding protein to a premature termination codon inhibits mammalian nonsense mediated mRNA decay, RNA (2008).
[69] G. Singh, I. Rebbapragada, J. Lykke-Andersen, A competition between stimulators and antagonists of Upf complex recruitment governs human nonsense-mediated mRNA decay, PLoS Biol. 6 (2008) e111.
[70] E. Conti, E. Izaurralde, Nonsense-mediated mRNA decay: molecular insights and mechanistic variations across species, Curr. Opin. Cell Biol. 17 (2005) 316–325.
[71] F. Lejeune, L.E. Maquat, Mechanistic links between nonsense-mediated mRNA decay and pre-mRNA splicing in mammalian cells, Curr. Opin. Cell Biol. 17 (2005) 309–315.
[72] S. Kertesz, Z. Kerenyi, Z. Merai, I. Bartos, T. Palfy, E. Barta, D. Silhavy, Both introns and long 3′-UTRs operate as cis-acting elements to trigger nonsense-mediated decay in plants, Nucleic Acids Res. 34 (2006) 6147–6157.
[73] A.V. Kochetov, A. Sarai, I.B. Rogozin, V.K. Shumny, L.L. Kolchanov, The role of alternative translation start sites in the generation of human protein diversity, Mol. Genet. Genomics 273 (2005) 491–496.
[73] A.M. Schwartz, T.V. Komarova, M.V. Skulachev, A.S. Zvereva, Y.L. Dorokhov, J.G. Atabekov, Stability of plant mRNAs depends on the length of the 3-untranslated region, Biochemistry (Mosc). 71 (2006) 1377–1384.
[74] P. Hilleren, R. Parker, mRNA surveillance in eukaryotes: kinetic proofreading of proper translation termination as assessed by mRNP domain organization? RNA 5 (1999) 711–719.
[75] R. Parker, H. Song, The enzymes and control of eukaryotic mRNA turnover, Nat. Struct. Mol. Biol. 11 (2004) 121–127.
[76] N.L. Garneau, J. Wilusz, C.J. Wilusz, The highways and byways of mRNA decay, Nat. Rev. Mol. Cell Biol. 8 (2007) 113–126.
[77] J. Lykke-Andersen, Identification of a human decapping complex associated with hUpf proteins in nonsense-mediated mRNA decay, Mol. Cell. Biol. 22 (2002) 8114–8121.
[78] F. Lejeune, X. Li, L.E. Maquat, Nonsense-mediated mRNA decay in mammalian cells involves decapping, deadenylating, and exonucleolytic activities, Mol. Cell 12 (2003) 675–687.
[79] D. Gatfield, E. Izaurralde, Nonsense-mediated messenger RNA decay is initiated by endonucleolytic cleavage in Drosophila, Nature 429 (2004) 575–578.
[80] T.O. Tange, A. Nott, M.J. Moore, The ever-increasing complexities of the exon junction complex, Curr. Opin. Cell Biol. 16 (2004) 279–284.
[81] C.B. Andersen, L. Ballut, J.S. Johansen, H. Chamieh, K.H. Nielsen, C.L. Oliveira, J.S. Pedersen, B. Seraphin, H. Le Hir, G.R. Andersen, Structure of the exon junction core complex with a trapped DEAD-box ATPase bound to RNA, Science 313 (2006) 1968–1972.
[82] F. Bono, J. Ebert, E. Lorentzen, E. Conti, The crystal structure of the exon junction complex reveals how it maintains a stable grip on mRNA, Cell 126 (2006) 713–725.
[83] T. Shibuya, T.O. Tange, N. Sonenberg, M.J. Moore, eIF4AIII binds spliced mRNA in the exon junction complex and is essential for nonsense-mediated decay, Nat. Struct. Mol. Biol. 11 (2004) 346–351.
[84] I.M. Palacios, D. Gatfield, D. St Johnston, E. Izaurralde, An eIF4AIII-containing complex required for mRNA localization and nonsense-mediated mRNA decay, Nature 427 (2004) 753–757.
[85] N.H. Gehring, G. Neu-Yilik, T. Schell, M.W. Hentze, A.E. Kulozik, Y14 and hUpf3b form an NMD-activating complex, Mol. Cell 11 (2003) 939–949.
[86] M.A. Ferraiuolo, C.S. Lee, L.W. Ler, J.L. Hsu, M. Costa-Mattioli, M.J. Luo, R. Reed, N. Sonenberg, A nuclear translation-like factor eIF4AIII is recruited to the mRNA during splicing and functions in nonsense-mediated decay, Proc. Natl. Acad. Sci. U. S. A. 101 (2004) 4118–4123.
[87] J. Dostie, G. Dreyfuss, Translation is required to remove Y14 from mRNAs in the cytoplasm, Curr. Biol. 12 (2002) 1060–1067.
[88] F. Lejeune, Y. Ishigaki, X. Li, L.E. Maquat, The exon junction complex is detected on CBP80-bound but not eIF4E-bound mRNA in mammalian cells: dynamics of mRNP remodeling, EMBO J. 21 (2002) 3536–3545.
[89] M. Buhler, S. Steiner, F. Mohn, A. Paillusson, O. Muhlemann, EJC-independent degradation of nonsense immunoglobulin-mu mRNA depends on 3′ UTR length, Nat. Struct. Mol. Biol. 13 (2006) 462–464.
[90] Y. Ishigaki, X. Li, G. Serin, L.E. Maquat, Evidence for a pioneer round of mRNA translation: mRNAs subject to nonsense-mediated decay in mammalian cells are bound by CBP80 and CBP20, Cell 106 (2001) 607–617.
[91] D. Matsuda, N. Hosoda, Y.K. Kim, L.E. Maquat, Failsafe nonsense-mediated mRNA decay does not detectably target eIF4E-bound mRNA, Nat. Struct. Mol. Biol. 14 (2007) 974–979.
[92] S.Y. Chiu, F. Lejeune, A.C. Ranganathan, L.E. Maquat, The pioneer translation initiation complex is functionally distinct from but structurally overlaps with the steady-state translation initiation complex, Genes Dev. 18 (2004) 745–754.
[93] O. Isken, L.E. Maquat, Quality control of eukaryotic mRNA: safeguarding cells from abnormal mRNA function, Genes Dev. 21 (2007) 1833–1856.
[94] Y.F. Chang, J.S. Imam, M.F. Wilkinson, The nonsense-mediated decay RNA surveillance pathway, Annu. Rev. Biochem. 76 (2007) 15.11–15.24.
[95] N. Hosoda, Y.K. Kim, F. Lejeune, L.E. Maquat, CBP80 promotes interaction of Upf1 with Upf2 during nonsense-mediated mRNA decay in mammalian cells, Nat. Struct. Mol. Biol. 12 (2005) 893–901.
[96] A.B. Maderazo, J.P. Belk, F. He, A. Jacobson, Nonsense-containing mRNAs that accumulate in the absence of a functional nonsense-mediated mRNA decay pathway are destabilized rapidly upon its restitution, Mol. Cell. Biol. 23 (2003) 842–851.
[97] Q. Gao, B. Das, F. Sherman, A.G. Maquat, Cap-binding protein 1-mediated and eukaryotic translation initiation factor 4E-mediated pioneer rounds of translation in yeast, Proc. Natl. Acad. Sci. U. S. A. 102 (2005) 4258–4263.

[98] M.M. Metzstein, M.A. Krasnow, Functions of the nonsense-mediated mRNA decay pathway in Drosophila development, PLoS Genet. 2 (2006) e180.
[99] J. Rehwinkel, I. Letunic, J. Raes, P. Bork, E. Izaurralde, Nonsense-mediated mRNA decay factors act in concert to regulate common mRNA targets, RNA 11 (2005) 1530–1544.
[100] M. Yoine, T. Nishii, K. Nakamura, Arabidopsis UPF1 RNA helicase for nonsense-mediated mRNA decay is involved in seed size control and is essential for growth, Plant Cell Physiol. 47 (2006) 572–580.
[101] S.M. Medghalchi, P.A. Frischmeyer, J.T. Mendell, A.G. Kelly, A.M. Lawler, H.C. Dietz, Rent1, a trans-effector of nonsense-mediated mRNA decay, is essential for mammalian embryonic viability, Hum. Mol. Genet. 10 (2001) 99–105.
[102] M.J. Ruiz-Echevarria, S.W. Peltz, The RNA binding protein Pub1 modulates the stability of transcripts containing upstream open reading frames, Cell 101 (2000) 741–751.
[103] X.L. Sun, X.J. Li, P.M. Moriarty, T. Henics, J.P. LaDuca, L.E. Maquat, Nonsense-mediated decay of mRNA for the selenoprotein phospholipids hydroperoxide glutathione peroxidase is detectable in cultured cells but masked or inhibited in rat tissues, Mol. Biol. Cell 12 (2001) 1009–1017.
[104] J.T. Mendell, N.A. Sharifi, J.L. Meyers, F. Martinez-Murillo, H.C. Dietz, Nonsense surveillance regulates expression of diverse classes of mammalian transcripts and mutes genomic noise, Nat. Genet. 36 (2004) 1073–1078.
[105] F. He, X. Li, P. Spatrick, R. Casillo, S. Dong, A. Jacobson, Genome-wide analysis of mRNAs regulated by the nonsense-mediated and 5′ to 3′ mRNA decay pathways in yeast, Mol. Cell 12 (2003) 1439–1452.
[106] D. Muhlrad, R. Parker, Aberrant mRNAs with extended 3′ UTRs are substrates for rapid degradation by mRNA surveillance, RNA 5 (1999) 1299–1307.
[107] J. Wittmann, E.M. Hol, H.M. Jack, hUPF2 silencing identifies physiologic substrates of mammalian nonsense-mediated mRNA decay, Mol. Cell. Biol. 26 (2006) 1272–1287.
[108] C. Giorgi, G.W. Yeo, M.E. Stone, D.B. Katz, C. Burge, G. Turrigiano, M.J. Moore, The EJC factor eIF4AIII modulates synaptic strength and neuronal protein expression, Cell 130 (2007) 179–191.
[109] Q.M. Mitrovich, P. Anderson, Unproductively spliced ribosomal protein mRNAs are natural targets of mRNA surveillance in C. elegans, Genes Dev. 14 (2000) 2173–2184.
[110] M. Morrison, K.S. Harris, M.B. Roth, smg mutants affect the expression of alternatively spliced SR protein mRNAs in Caenorhabditis elegans, Proc. Natl. Acad. Sci. U. S. A. 94 (1997) 9782–9785.
[111] A. Sureau, R. Gattoni, Y. Dooghe, J. Stevenin, J. Soret, SC35 autoregulates its expression by promoting splicing events that destabilize its mRNAs, EMBO J. 20 (2001) 1785–1796.
[112] E.V. Makeyev, J. Zhang, M.A. Carrasco, T. Maniatis, The microRNA miR-124 promotes neuronal differentiation by triggering brain-specific alternative pre-mRNA splicing, Mol. Cell 27 (2007) 435–448.
[113] J.C. Schoning, C. Streitner, D.R. Page, S. Hennig, K. Uchida, E. Wolf, M. Furuya, D. Staiger, Auto-regulation of the circadian slave oscillator component At GRP7 and regulation of its targets is impaired by a single RNA recognition motif point mutation, Plant J. (2007).
[114] G. Neu-Yilik, N.H. Gehring, M.W. Hentze, A.E. Kulozik, Nonsense-mediated mRNA decay: from vacuum cleaner to Swiss army knife, Genome Biol. 5 (2004) 218.
[115] E. Donnadieu, M.H. Jouvin, S. Rana, M.F. Moffatt, E.H. Mockford, W.O. Cookson, J.P. Kinet, Competing functions encoded in the allergy-associated F(c)epsilon-RIbeta gene, Immunity 18 (2003) 665–674.
[116] B. Kebaara, T. Nazarenus, R. Taylor, A.L. Atkin, Genetic background affects relative nonsense mRNA accumulation in wild-type and upf mutant yeast strains, Curr. Genet. 43 (2003) 171–177.
[117] T.P. Kerr, C.A. Sewry, S.A. Robb, R.G. Roberts, Long mutant dystrophins and variable phenotypes: evasion of nonsense-mediated decay? Hum. Genet. 109 (2001) 402–407.
[118] J.F. Bateman, S. Freddi, G. Nattrass, R. Savarirayan, Tissue-specific RNA surveillance? Nonsense-mediated mRNA decay causes collagen X haploinsufficiency in Schmid metaphyseal chondrodysplasia cartilage, Hum. Mol. Genet. 12 (2003) 217–225.
[119] L. Linde, S. Boelz, G. Neu-Yilik, A.E. Kulozik, B. Kerem, The efficiency of nonsense-mediated mRNA decay is an inherent character and varies among different cells, Eur. J. Hum. Genet. (2007).
[120] N. Resta, F.C. Susca, M.C. Di Giacomo, A. Stella, N. Bukvic, R. Bagnulo, C. Simone, G. Guanti, A homozygous frameshift mutation in the ESCO2 gene: evidence of intertissue and interindividual variation in Nmd efficiency, J. Cell. Physiol. 209 (2006) 67–73.
[121] L. Linde, S. Boelz, M. Nissim-Rafinia, Y.S. Oren, M. Wilschanski, Y. Yaacov, D. Virgilis, G. Neu-Yilik, A.E. Kulozik, E. Kerem, B. Kerem, Nonsense-mediated mRNA decay affects nonsense transcript levels and governs response of cystic fibrosis patients to gentamicin, J. Clin. Invest. 117 (2007) 683–692.
[122] P.A. Frischmeyer, H.C. Dietz, Nonsense-mediated mRNA decay in health and disease, Hum. Mol. Genet. 8 (1999) 1893–1900.
[123] M.H. Viegas, N.H. Gehring, S. Breit, M.W. Hentze, A.E. Kulozik, The abundance of RNPS1, a protein component of the exon junction complex, can determine the variability in efficiency of the Nonsense Mediated Decay pathway, Nucleic Acids Res. 35 (2007) 4542–4551.
[124] L. Nott, H. Le Hir, M.J. Moore, Splicing enhances translation in mammalian cells: an additional function of the exon junction complex, Genes Dev. 18 (2004) 210–222.
[125] A.B. Maderazo, F. He, D.A. Mangus, A. Jacobson, Upf1p control of nonsensemRNA translation is regulated by Nmd2p and Upf3p, Mol. Cell. Biol. 20 (2000) 4591–4603.
[126] V. Lundblad, J.W. Szostak, A mutant with a defect in telomere elongation leads to senescence in yeast, Cell 57 (1989) 633–643.

[127] P. Reichenbach, M. Hoss, C.M. Azzalin, M. Nabholz, P. Bucher, J. Lingner, A human homolog of yeast est1 associates with telomerase and uncaps chromosome ends when overexpressed, Curr. Biol. 13 (2003) 568–574.
[128] B.E. Snow, N. Erdmann, J. Cruickshank, H. Goldman, R.M. Gill, M.O. Robinson, L. Harrington, Functional conservation of the telomerase protein Est1p in humans, Curr. Biol. 13 (2003) 698–704.
[129] C.M. Azzalin, P. Reichenbach, L. Khoriauli, E. Giulotto, J. Lingner, Telomeric repeat containing RNA and RNA surveillance factors at mammalian chromosome ends, Science 318 (2007) 798–801.
[130] K.M. Brumbaugh, D.M. Otterness, C. Geisen, V. Oliveira, J. Brognard, X. Li, F. Lejeune, R.S. Tibbetts, L.E. Maquat, R.T. Abraham, The mRNA surveillance protein hSMG-1 functions in genotoxic stress response pathways in mammalian cells, Mol. Cell 14 (2004) 585–598.
[131] C.M. Azzalin, J. Lingner, The human RNA surveillance factor UPF1 is required for S phase progression and genome stability, Curr. Biol. 16 (2006) 433–439.
[132] L.M. Carastro, C.K. Tan, M. Selg, H.M. Jack, A.G. So, K.M. Downey, Identification of delta helicase as the bovine homolog of HUPF1: demonstration of an interaction with the third subunit of DNA polymerase delta, Nucleic Acids Res. 30 (2002) 2232–2243.
[133] M.E. Domeier, D.P. Morse, S.W. Knight, M. Portereiko, B.L. Bass, S.E. Mango, A link between RNA interference and nonsense-mediated decay in Caenorhabditis elegans, Science 289 (2000) 1928–1931.
[134] J. Hock, L. Weinmann, C. Ender, S. Rudel, E. Kremmer, M. Raabe, H. Urlaub, G. Meister, Proteomic and functional analysis of Argonaute-containing mRNA-protein complexes in human cells, EMBO Rep. 8 (2007) 1052–1060.
[135] J.E. Dahlberg, E. Lund, E.B. Goodwin, Nuclear translation: what is the evidence? RNA 9 (2003) 1–8.
[136] M.F. Wilkinson, A.B. Shyu, RNA surveillance by nuclear scanning? Nat. Cell Biol. 4 (2002) E144–147.
[137] F.J. Iborra, D.A. Jackson, P.R. Cook, Coupled transcription and translation within nuclei of mammalian cells, Science 293 (2001) 1139–1142.
[138] M. Buhler, M.F. Wilkinson, O. Muhlemann, Intranuclear degradation of nonsense codon-containing mRNA, EMBO. Rep. 3 (2002) 646–651.
[139] G. Singh, S. Jakob, M.G. Kleedehn, J. Lykke-Andersen, Communication with the exon-junction complex and activation of nonsense-mediated decay by human Upf proteins occur in the cytoplasm, Mol. Cell 27 (2007) 780–792.
[140] U. Sheth, R. Parker, Targeting of aberrant mRNAs to cytoplasmic processing bodies, Cell 125 (2006) 1095–1109.
[141] A. Eulalio, I. Behm-Ansmant, E. Izaurralde, P bodies: at the crossroads of post-transcriptional pathways, Nat. Rev., Mol. Cell Biol. 8 (2007) 9–22.
[142] D.A. Bushner, M.F. Trudeau, M.R. Meister, SCNM1, a putative RNA splicing factor that modifies disease severity in mice, Science 301 (2003) 967–969.
[143] H.X. Liu, L. Cartegni, M.Q. Zhang, A.R. Krainer, A mechanism for exon skipping caused by nonsense or missense mutations in BRCA1 and other genes, Nat. Genet. 27 (2001) 55–58.
[144] F. Pagani, E. Buratti, C. Stuani, F.E. Baralle, Missense, nonsense, and neutral mutations define juxtaposed regulatory elements of splicing in cystic fibrosis transmembrane regulator exon 9, J. Biol. Chem. 278 (2003) 26580–26588.
[145] M. Buhler, O. Muhlemann, Alternative splicing induced by nonsense mutations in the immunoglobulin mu VDJ exon is independent of truncation of the open reading frame, RNA 11 (2005) 139–146.
[146] M. Caputi, R.J. Kendzior, Jr., K.L. Beemon, A nonsense mutation in the fibrillin-1 gene of a Marfan syndrome patient induces NMD and disrupts an exonic splicing enhancer, Genes Dev. 16 (2002) 1754–1759.
[147] P. Mohn, M. Buhler, O. Muhlemann, Nonsense-associated alternative splicing of T-cell receptor beta genes: no evidence for frame dependence, RNA 11 (2005) 147–156.
[148] H.C. Dietz, R.J. Kendzior, Jr., Maintenance of an open reading frame as an additional level of scrutiny during splice site selection, Nat. Genet. 8 (1994) 183–188.
[149] B. Li, C. Wachtel, E. Miriami, G. Yahalom, G. Friedlander, G. Sharon, R. Sperling, J. Sperling, Stop codons affect 5′ splice site selection by surveillance of splicing, Proc. Natl. Acad. Sci. U. S. A. 99 (2002) 5277–5282.
[150] O. Muhlemann, C.S. Mock-Casagrande, J. Wang, S. Li, N. Custodio, M. Carmo-Fonseca, M.F. Wilkinson, M.J. Moore, Precursor RNAs harboring nonsense codons accumulate near the site of transcription, Mol. Cell 8 (2001) 33–43.
[151] J. Wang, J.I. Hamilton, M.S. Carter, S. Li, M.F. Wilkinson, Alternatively spliced TCR mRNA induced by disruption of reading frame, Science 297 (2002) 108–110.
[152] Y.F. Chang, W.K. Chan, J.S. Imam, M.F. Wilkinson, Alternatively spliced T-cell receptor transcripts are up-regulated in central relative to peripheral lymphoid tissues, J. Biol. Chem. 282 (2007) 29738–29747.
[153] M. Buhler, F. Mohn, L. Stalder, O. Muhlemann, Transcriptional silencing of nonsense codon-containing immunoglobulin minigenes, Mol. Cell 18 (2005) 307–317.
[154] L. Stalder, O. Muhlemann, Transcriptional silencing of nonsense codon-containing immunoglobulin mini-genes requires translation of its mRNA, J. Biol. Chem. 282 (2007) 16079–16085.
[155] H.A. Kuzmiak, L.E. Maquat, Applying nonsense-mediated mRNA decay research to the clinic: progress and challenges, Trends Mol. Med. 12 (2006) 306–316.
[156] G.W. Hall, S. Thein, Nonsense codon mutations in the terminal exon of the beta-globin gene are not associated with a reduction in beta-mRNA accumulation: a mechanism for the phenotype of dominant beta-thalassemia, Blood 83 (1994) 2031–2037.
[157] K. Inoue, M. Khajavi, T. Ohyama, S. Hirabayashi, J. Wilson, J.D. Reggin, P. Mancias, I. J. Butler, M.F. Wilkinson, M. Wegner, J.R. Lupski, Molecular mechanism for distinct neurological phenotypes conveyed by allelic truncating mutations, Nat. Genet. 36 (2004) 361–369.

[158] T. Hermann, Aminoglycoside antibiotics: old drugs and new therapeutic approaches, Cell. Mol. Life Sci. 64 (2007) 1841–1852.
[159] R. Kellermayer, Translational readthrough induction of pathogenic nonsense mutations, Eur. J. Med. Genet. 49 (2006) 445–450.
[160] M. Wilschanski, Y. Yahav, Y. Yaacov, H. Blau, L. Bentur, J. Rivlin, M. Aviram, T. Bdolah-Abram, Z. Bebok, L. Shushi, B. Kerem, E. Kerem, Gentamicin-induced correction of CFTR function in patients with cystic fibrosis and CFTR stop mutations, N. Engl. J. Med. 349 (2003) 1433–1441.
[161] E.M. Welch, E.R. Barton, J. Zhuo, Y. Tomizawa, W.J. Friesen, P. Trifillis, S. Paushkin, M. Patel, C.R. Trotta, S. Hwang, R.G. Wilde, G. Karp, J. Takasugi, G. Chen, S. Jones, H. Ren, Y.C. Moon, D. Corson, A.A. Turpoff, J.A. Campbell, M. Conn, A. Khan, N.G. Almstead, J. Hedrick, A. Mollin, N. Risher, M. Weetall, S. Yeh, A.A. Branstrom, J.M. Colacino, J. Babiak, W.D. Ju, S. Hirawat, V.J. Northcutt, L.L. Miller, P. Spatrick, F. He, M. Kawana, H. Feng, A. Jacobson, S.W. Peltz, PTC124 targets genetic disorders caused by nonsense mutations, Nature 447 (2007) 87–91.

6.3.6. Paper VI

Stalder L, Mühlemann O. Processing bodies are not required for mammalian NMD. RNA. 2009 Jul;15(7):1265-73.

This publication can not be printed here due to copyright protection. Please download this publication from http://rnajournal.cshlp.org

Die VDM Verlagsservicegesellschaft sucht für wissenschaftliche Verlage abgeschlossene und herausragende

Dissertationen, Habilitationen, Diplomarbeiten, Master Theses, Magisterarbeiten usw.

für die kostenlose Publikation als Fachbuch.

Sie verfügen über eine Arbeit, die hohen inhaltlichen und formalen Ansprüchen genügt, und haben Interesse an einer honorarvergüteten Publikation?

Dann senden Sie bitte erste Informationen über sich und Ihre Arbeit per Email an *info@vdm-vsg.de*.

Sie erhalten kurzfristig unser Feedback!

VDM Verlagsservicegesellschaft mbH
Dudweiler Landstr. 99
D - 66123 Saarbrücken
www.vdm-vsg.de

Telefon +49 681 3720 174
Fax   +49 681 3720 1749

Die VDM Verlagsservicegesellschaft mbH vertritt

Printed by Books on Demand GmbH, Norderstedt / Germany